Progress in Probability and Statistics
Volume 16

Series Editor
Murray Rosenblatt

Yuri Kifer

Random Perturbations of Dynamical Systems

1988

Birkhäuser
Boston · Basel

Yuri Kifer
Institute of Mathematics
The Hebrew University of Jerusalem
Givat Ram, Jerusalem
Israel
and
Department of Mathematics
Cornell University
Ithaca, NY 14853
U.S.A.

Library of Congress Cataloging-in-Publication Data
Kifer, Yuri, 1948–
Random perturbations of dynamical systems.
(Progress in probability and statistics ; v. 16)
Bibliography: p.
Includes index.
1. Stochastic processes. 2. Perturbations (Mathematics) 3. Differentiable dynamical systems. I. Title. II. Series.
QA274.K55 1988 519.2 87-38199
ISBN 0-8176-3384-7

CIP-Titelaufnahme der Deutschen Bibliothek
Kifer, Yuri:
Random perturbations of dynamical systems / Yuri Kifer.—
Boston ; Basel : Birkhäuser, 1988
(Progress in probability and statistics ; Vol. 16)
ISBN 978-1-4615-8183-3 ISBN 978-1-4615-8181-9 (eBook)
DOI 10.1007/978-1-4615-8181-9
NE: GT

Softcover reprint of the hardcover 1st edition 1988

ISBN 978-1-4615-8183-3

Text prepared by the author in camera-ready form.

9 8 7 6 5 4 3 2 1

Frequently used notations

$C(M)$ - the space of continuous functions on M

C^k-class - continuous together with k-derivatives

D - the differential of a map F

Er - the expectation of a random variable r

F^n - iterates of a map F; F^t - a flow

$h_\rho(F)$ - the metric entropy of F with respect to an invariant measure ρ

χ_A - the indicator of a set A, i.e., $\chi_A(x) = 1$ if $x \in A$ and $= 0$, for otherwise

$\mathcal{P}(M)$ - the space of Borel probability measures on M

$P\{A\}$ - the probability of an event A

$P^\epsilon(x,\cdot)$, $P^\epsilon(n,x,\cdot)$, $P^\epsilon(t,x,\cdot)$ - transition probabilities of Markov chains X_n^ϵ, Markov processes X_t^ϵ

P^ϵ, P_t^ϵ - corresponding operators acting on functions and $(P^\epsilon)^*$, $(P_t^\epsilon)^*$-adjoint operators acting on measures

$\mathbb{R}^\nu$ - the ν-dimensional Euclidean space

TM - the tangent bundle of a smooth manifold M

X_n^ϵ, X_t^ϵ - Markov chains, Markov processes which are random perturbations of iterates of a map F, of a flow F^t

□ - the end of a proof

Statement i, j - i denotes the section and j denotes the number of this statement in the section. Propositions and Theorems have uniform numeration. Lemmas, Corollaries, Examples, Remarks, and Figures have their own numerations. The Roman number at the beginning (for instance, III.1.2) means the number of the chapter.

Table of Contents

Introduction 1

I. General analysis of random perturbations 7

1.1. Convergence of invariant measures 7

1.2. Entropy via random perturbations: generalities 30

1.3. Locating invariant sets 39

1.4. Attractors and limiting measures 43

1.5. Attractors and limiting measures via large deviations 56

II. Random perturbations of hyperbolic and expanding transformations 92

2.1. Preliminaries 92

2.2. Markov chains in tangent bundles 106

2.3. Hyperbolic and expanding transformations 123

2.4. Limiting measures 141

2.5. Sinai-Bowen-Ruelle's measures. Discussion. 155

2.6. Entropy via random perturbations 165

2.7. Stability of the topological pressure 178

2.8. Appendix: proof of (1.12) 189

III. Applications to partial differential equations 197

3.1. Principal eigenvalue and invariant sets 197

3.2. Localization theorem 211

3.3. Random perturbations and spectrum 231

IV. Random perturbations of some special models 252

4.1. Random perturbations of one-dimensional transformations 252

4.2. Misiurewicz's maps of an interval 270

4.3. Lorenz's type models 275

Bibliography 283

Index 293

Table of Contents

Introduction

Mathematicians often face the question to which extent mathematical models describe processes of the real world. These models are derived from experimental data, hence they describe real phenomena only approximately. Thus a mathematical approach must begin with choosing properties which are not very sensitive to small changes in the model, and so may be viewed as properties of the real process. In particular, this concerns real processes which can be described by means of ordinary differential equations. By this reason different notions of stability played an important role in the qualitative theory of ordinary differential equations commonly known nowdays as the theory of dynamical systems. Since physical processes are usually affected by an enormous number of small external fluctuations whose resulting action would be natural to consider as random, the stability of dynamical systems with respect to random perturbations comes into the picture. There are differences between the study of stability properties of single trajectories, i.e., the Lyapunov stability, and the global stability of dynamical systems. The stochastic Lyapunov stability was dealt with in Hasminskii [Has].

In this book we are concerned mainly with questions of global stability in the presence of noise which can be described as recovering parameters of dynamical systems from the study of their random perturbations. The parameters which is possible to obtain in this way can be considered as stable under random perturbations, and so having physical sense.

Our set up is the following. We consider a metric space M together with a continuous map $F: M \to M$. By random perturbations of F we mean a family of Markov chains X_n^ϵ whose transition probabilities $P^\epsilon(x,\Gamma) = P\{X_{n+1}^\epsilon \in \Gamma | X_n^\epsilon = x\}$ converge in some sense as $\epsilon \to 0$ to the unit mass at Fx. This means that due to random fluctuations a particle misses the point Fx and falls in a random point whose distribution is close to the δ-function at Fx. One obtains important partial cases of this situation when the particle falls in ϵ-neighborhood of Fx (local random perturbations) or when after jumping to Fx the particle performs a diffusion for the time ϵ (diffusion type random perturbations). The last model has a continuous time counterpart which represents diffusion processes with a small parameter in diffusion terms. Assuming that random perturbations are caused by a large number of small independent random fluctuations one can deduce the legitimacy of diffusion type random perturbations via an appropriate version of the central limit theorem. Remark that in the continuous time case this is the only available continuous in time smooth model.

Markov chains X_n^ϵ may have invariant measures μ^ϵ. The study of the asymptotic behavior of μ^ϵ as $\epsilon \to 0$ will be one of the main goals of this book. An assertion obtained originally by Khasminskii [Kh] says that any weak limit as $\epsilon \to 0$ of measures μ^ϵ must be an invariant measure of the map F. A natural question arising here is how to describe these limiting measures, and when μ^ϵ converges as $\epsilon \to 0$ to a single measure μ which can be viewed then as most stable to random perturbations. The importance of such measures was underlined by Ruelle [Ru5]-[Ru7] in connection with mathematical models for the phenomenon of hydrodinamic turbulence. Dynamical systems involved in these models have very complicated structure and possess a wealth of invariant measures. It is natural to assume that physically relevant measures which may describe turbulence must be stable to random perturbations.

The problem of what happens to stationary distribution of a random process arising as a result of random perturbations of a dynamical system when these perturbations decrease has been studied for the first time by Pontrjagin, Andronov and Vitt [PAV] in 1933 who considered a one-dimensional process with small diffusion. The problem was promoted by Kolmogorov in the fifties and sixties leading to a number of papers by Khasminskii [Kh] and Wentzell and Freidlin [WF] who dealt with diffusion type random perturbations of relatively simple from the dynamical point of view systems. The first approach to random perturbations of systems with complicated dynamics was proposed by Sinai [Si1] who suggested the problem to the author of this book whose first work [Ki1] on this subject appeared in 1974 which was later generalized in [Ki10]. In these papers the problem was solved for a wide class of hyperbolic dynamical systems.

Our exposition in this book concerns mainly (except for Chapter III) random perturbations of a single map F, i.e., we consider the discrete time case. This leads to a substantial generalization and, at the same time, simplification since the only workable continuous time model involving diffusion type random perturbations requires from the reader some knowledge of diffusion processes and partial differential equations. As a result probabilistic prerequisites are rather modest and only some acquaintance with Markov chains is needed. On the other hand, we shall obtain the results for diffusion type perturbations as a partial case. Our exposition takes care about probabilistic audience, as well, so that most parts of the book do not require special dynamical systems prerequisities.

There is Freidlin and Wentzell's [FW] book with the same title as the present one which studies effects of large deviations type for diffusion perturbations of systems with usually relatively simple dynamics. The outcome in their study depends on perturbations (diffusion coefficients), and so it has no connection with questions of stability of dynamical systems. We shall consider mainly the situation where properties of random perturbations (which are rather general) will be influenced decisively by the complicated dynamics of the deterministic motion and they will not depend essentially on perturbations. As a result our book has no connections with Freidlin and Wentzell's one except for our Section 1.5 in which we generalize results of Chapter 6 from their book claiming that limiting measures of random perturbations sit on attractors of the corresponding dynamical system.

This book has the following structure. Chapter I deals with general properties of random perturbations which do not involve essentially a study of delicate dynamics of corresponding systems. In Section 1.1 we give necessary definitions, prove preliminary results about convergence of invariant measures of random perturbations, and study conditions which ensure the existence of invariant measures for Markov chains. Sections 1.2 and 1.3 explain how one can obtain some information about the entropy and invariant sets of dynamical systems via their random perturbations. In Sections 1.4 and 1.5 we exhibit conditions which ensure that all weak limits as $\epsilon \to 0$ of invariant mesures of random perturbations X_n^ϵ of a map F have support on attractors of F. In Section 1.4 we follow Ruelle [Ru5] to prove this result for localized perturbations and in Section 1.5 we generalize Wentzell and Freidlin's approach to derive the result for random perturbations satisfying certain large deviations condition.

In Chapter II we study random perturbations of dynamical systems with some hyperbolicity or expanding properties. We prove here results concerning the convergence of invariant measures of random perturbations, and show how the entropy and the topological pressure can be obtained via random perturbations. Though this chapter is close to the author's paper [Ki10] we revise the exposition in such a way that in most parts we rely only on qualitative properties of dynamical systems under consideration, collected in Section 2.3. Hard questions of ergodic theory are left for a discussion in Section 2.5 and they are not used in proofs. This approach enables us to emphasize properties needed for each step of the proof and together with a number of examples it makes possible for nonexperts in dynamical systems to follow the exposition.

In Chapter III we study diffusion perturbations of continuous time dynamical systems, i.e., flows, and apply the results to partial differential equations. We shall study there the asymptotical behavior of the principal eigenvalue for generators of diffusions in compact domains which are random perturbations of the corresponding flows. It turns out that the outcome will depend decisively on invariant sets the flow has in a domain. In the last section we determine the asymptotical behavior of the whole spectrum for generators of diffusion perturbations of a constant vector field on a torus, and discuss the corresponding problem in the general case. Our exposition follows mainly the papers [Ki4], [Ki6], [Ki8] and [EK].

Chapter IV deals with random perturbations of dynamical systems which are not structurally stable and do not satisfy precise hyperbolicity or expanding conditions. In Section 4.1 we consider an expanding map of an interval with singularities and in Section 4.2 we study random perturbations of Misiurewicz's transformations of an interval which are known to be unstable with respect to deterministic perturbations. In Section 4.3 we discuss random perturbations of Lorenz's type model dynamical systems which are not structurally stable and, nevertheless, possess certain stability under random perturbations. These show both the flexibility of our methods and the importance of the study of random perturbations of dynamical systems which may extend our understanding of stability properties of dynamical systems. The exposition in Chapter IV is less detailed than in previous chapters and in many places we give only ideas of the proof or refer the reader to other papers.

All chapters are meant to be read in order except that Sections 1.5, 2.8, 3.3, and Chapter IV may be omitted in the first reading.

Some of the results in this book have not been yet published at all, others have appeared only in the periodic literature. The theory of random perturbations of dynamical systems is just being created, it did not take yet its final form, and there is still much to be done.

This book is addressed to mathematicians and mathematical physicists working in probability and (or) dynamical systems, and can be read also by graduate students with some background in these areas.

During the work on this book the author was supported by United States-Israel Binational Science Foundation Grant #84-00028.

The final phase of the work on this book was done during the author's visit to the Department of Mathematics of Cornell University when he was also partly supported by the U.S. Army Research Office through the Mathematical Sciences Institute of Cornell University. The camera-ready manuscript was prepared by the proficient typing staff of the Department of Mathematics and MSI of Cornell University.

Chapter I

General analysis of random perturbations

In this chapter we study the asymptotic behavior of random perturbations of dynamical systems in rather general circumstances.

1.1. Convergence of invariant measures.

In this section we shall define random perturbations and study their basic properties.

We start with a metric space M and a continuous map $F: M \to M$. Let $\mathcal{P}(M)$ denote the space of Borel probability measures on M with the topology of weak convergence. Consider a family $Q_x^\epsilon \in \mathcal{P}(M)$ defined for any $x \in M$ and each $\epsilon > 0$ such that all maps $Q_\cdot^\epsilon : M \to \mathcal{P}(M)$ sending x to Q_x^ϵ are Borel. In addition, we shall assume that for each bounded continuous function g on M,

$$\lim_{\epsilon \to 0} \sup_{x \in M} \left| \int_M g(y) Q_x^\epsilon(dy) - g(x) \right| = 0. \tag{1.1}$$

A family of Markov chains X_n^ϵ, $n = 0,1,2,\ldots$ with transition probabilities

$$P^\epsilon(x,\Gamma) = P\{X_{n+1}^\epsilon \in \Gamma \mid X_n^\epsilon = x\} = Q_{Fx}^\epsilon(\Gamma) \tag{1.2}$$

defined for any $x \in M$ and a Borel set $\Gamma \subset M$, will be called small random perturbations of the transformation F. The meaning is that a particle jumps from x to Fx and then disperses randomly near Fx with the distribution

Q^ϵ_{Fx}. Another interpretation says that due to a random error the particle misses the point Fx and falls in a random point whose distribution is close to the δ-function at Fx. Remark that we consider all Markov chains X^ϵ_n on the same probability space $(\Omega, \mathcal{F}, P)$.

We shall say that a probability measure μ^ϵ on M is an invariant measure of the Markov chain X^ϵ_n if for any Borel set $\Gamma \subset M$,

$$\int_M d\mu^\epsilon(x) P^\epsilon(x, \Gamma) = \mu^\epsilon(\Gamma). \tag{1.3}$$

As usual, for $\nu_i, \nu \in \mathcal{P}(M)$, $i = 1, 2, \ldots$ we shall say that $\nu_i \to \nu$ in the weak sense (and write $\nu_i \xrightarrow{w} \nu$) if $\int g d\nu_i \to \int g d\nu$ as $i \to \infty$ for any bounded function $g \in C(M)$ where $C(M)$ denotes the space of continuous functions on M. The following simple fact established initially by Khasminskii [Kh] is a starting point for our study.

Theorem 1.1. *Suppose that (1.1)-(1.3) are satisfied and*

$$\mu^{\epsilon_i} \xrightarrow{w} \mu \tag{1.4}$$

for some subsequence $\epsilon_i \to 0$. *Then* μ *is an invariant measure of the map* F *(F-invariant measure), i.e., for any Borel set* $\Gamma \subset M$,

$$\mu(F^{-1}\Gamma) = \mu(\Gamma). \tag{1.5}$$

Proof. Take a bounded continuous function g then by (1.3),

$$\int_M g d\mu^\epsilon = \int_M \int_M d\mu^\epsilon(x) P^\epsilon(x, dy) g(y). \tag{1.6}$$

Hence by (1.2),

$$\left|\int_M g(x)d\mu(x) - \int_M g(Fx)d\mu(x)\right|$$

$$\leq \left|\int_M g(x)d\mu(x) - \int_M g(x)d\mu^{\epsilon}(x)\right|$$

$$+ \left|\int_M\left(\int_M Q^{\epsilon}_{Fx}(dy)g(y) - g(Fx)\right)d\mu^{\epsilon}(x)\right|$$

$$+ \left|\int_M g(Fx)d\mu^{\epsilon}(x) - \int_M g(Fx)d\mu(x)\right|.$$

Letting here $\epsilon \to 0$ along the subsequence ϵ_i we obtain by (1.1) and (1.4) that

$$\int_M g(x)d\mu(x) = \int_M g(Fx)d\mu(x). \tag{1.7}$$

The relation (1.7) being true for any bounded $g \in C(M)$ implies (1.5). □

Define "vague" random perturbations requiring (1.1) to be true only for continuous functions g having compact supports. Introduce also the following condition: for any $\delta > 0$,

$$\lim_{\epsilon\to 0} \sup_{x\in M} Q^{\epsilon}_{x}(M\backslash U_{\delta}(x)) = 0 \tag{1.8}$$

where $U_{\delta}(x) = \{y:\mathrm{dist}(x,y) < \delta\}$.

Theorem 1.2. (i) *If (1.1) holds true for any bounded $g \in C(M)$, then (1.8) is also satisfied provided M is a complete metric space;*

(ii) *If (1.8) is satisfied, then (1.1) holds true for any $g \in C(M)$ having a compact support;*

(iii) *Let M be a locally compact space and let measures μ^{ϵ} satisfy (1.3). Suppose that (1.1) is true for any $g \in C(M)$ with a compact support and for such* *functions* $\int g d\mu^{\epsilon_i} \to \int g d\mu$, *i.e.*, $\mu^{\epsilon_i} \in \mathcal{P}(M)$ *converges vaguely to* $\mu \in \mathcal{P}(M)$. *Then* μ *is F-invariant. This means that for a locally compact space* M, *Theorem 1.1 remains true for vague random perturbations.*

Proof. (i) Suppose that (1.8) fails, though (1.1) is true. Then there exist numbers $\gamma, \delta > 0$, a sequence of points $x_1, x_2, \ldots \in M$ and a sequence of numbers $\epsilon_i > 0$, $\epsilon_i \to 0$ such that

$$Q_{x_i}^{\epsilon_i}(U_\delta(x_i)) < 1-\gamma, \tag{1.9}$$

for all $i = 1,2,\ldots$. Suppose that the sequence $x_1, x_2, \ldots$ has a limit point $x_0 \in M$, i.e., $x_{i_j} \to x_0$ as $j \to \infty$ for a subsequence x_{i_j}. Take a continuous function g such that $0 \le g \le 1$, $g(x) = 1$ for $x \in U_{\delta/3}(x_0)$ and $g(x) = 0$ for $x \notin U_{2\delta/3}(x_0)$. Then by (1.9) for all j big enough,

$$|g(x_{i_j}) - \int_M Q_{x_{i_j}}^{\epsilon_{i_j}}(dy)g(y)| = \int_M (1-g(x_{i_j}))Q_{x_{i_j}}^{\epsilon_{i_j}}(dy)$$

$$\ge Q_{x_{i_j}}^{\epsilon_{i_j}}(M \backslash U_\delta(x_{i_j})) > \gamma$$

which contradicts (1.1) since $\epsilon_i \to 0$.

Now let the sequence x_i have no limit points. Since M is complete this means that if $\delta > 0$ is small enough one can pass, maybe, to a subsequence to obtain that $\mathrm{dist}(x_i, x_j) > \delta$ for all $i, j \ge 1$ provided $i \ne j$. Assume, again that (1.9) holds true. Notice that (1.1) implies

$$Q_{x_i}^{\epsilon_i}\left(\bigcup_{j=1}^{k} U_\delta(x_j)\right) \to 0 \quad \text{as} \quad i \to \infty \tag{1.10}$$

for each fixed $k \geq 1$. Indeed, if (1.10) were not true then for some subsequence i_j one has

$$Q_{x_{i_\ell}}^{\epsilon_{i_\ell}}\left(\bigcup_{j=1}^{k} U_\delta(x_j)\right) > \eta > 0 \quad \text{for all} \quad \ell.$$

Pick up a continuous function g such that $0 \leq g \leq 1$, $g(x) = 1$ for $x \in \bigcup_{\ell=k}^{\infty} U_{\delta/3}(x_{i_\ell})$ and $g(x) = 0$ if $x \notin \bigcup_{\ell=k}^{\infty} U_{2\delta/3}(x_{i_\ell})$. Then for all $\ell > k$,

$$|g(x_{i_\ell}) - \int_M Q_{x_{i_\ell}}^{\epsilon_{i_\ell}}(dy)g(y)| = \int_M (1-g(x_{i_\ell}))Q_{x_{i_\ell}}^{\epsilon_{i_\ell}}(dy)$$

$$\geq Q_{x_{i_\ell}}^{\epsilon_{i_\ell}}\left(\bigcup_{j=1}^{k} U_\delta(x_j)\right) > \eta$$

which contradicts (1.1) since $\epsilon_{i_\ell} \to 0$. Finally, using (1.10) and the fact that for each i,

$$Q_{x_i}^{\epsilon_i}\left(\bigcup_{\ell=k}^{\infty} U_\delta(x_\ell)\right) \to 0 \quad \text{as} \quad k \to \infty$$

we can choose a subsequence i_j such that

$$Q_{x_{i_j}}^{\epsilon_{i_j}}\left(\bigcup_{n:n\neq j} U_\delta(x_{i_n}\right) < \gamma/2 \tag{1.11}$$

for all $j = 1,2,\ldots$. Define a continuous function g such that $0 \leq g \leq 1$, $g(x) = 1$ for $x \in \bigcup_{j=1}^{\infty} U_{\delta/3}(x_{i_j})$ and

$g(x) = 0$ if $x \notin \bigcup_{j=1}^{\infty} U_{2\delta/3}(x_{i_j})$. Then by (1.9) and (1.11) for all j,

$$|g(x_{i_j}) - \int_M Q_{x_{i_j}}^{\epsilon_{i_j}}(dy)g(y)| = \int_M (1-g(x_{i_j}))Q_{x_{i_j}}^{\epsilon_{i_j}}(dy)$$

$$\geq Q_{x_{i_j}}^{\epsilon_{i_j}}(M\setminus \bigcup_{n=1}^{\infty} U_{\delta}(x_{i_n})) > \gamma/2$$

which contradicts (1.1) since $\epsilon_{i_j} \to 0$.

(ii) If $g \in C(M)$ has a compact support then g is bounded and uniformly continuous. Thus (1.8) implies

$$|\int_M g(y)Q_x^{\epsilon}(dy) - g(x)|$$

$$\leq \sup_{y\in U_{\delta}(x)} |g(y)-g(x)| + 2 \sup_y |g(y)| Q_x^{\epsilon}(M\setminus U_{\delta}(x)) \to 0$$

uniformly in x when, first, $\epsilon \to 0$ and then $\delta \to 0$.

(iii) Let $g \in C(M)$ has a compact support. Then in the same way as in the proof of Theorem 1.1 we conclude that (1.3) implies (1.6) which, in turn, yields (1.7). Since in a locally compact space M compact sets generate the Borel σ-field then (1.7) being true for any $g \in C(M)$ with a compact support implies (1.5). □

Remark 1.1. By analogy with vague random perturbations which correspond to the vague convergence we could refer to random perturbations defined at the beginning by (1.1) as to weak random perturbations since they correspond to the weak convergence in Theorem 1.1. Nevertheless, since we shall deal mainly with the latter we shall call them simply random perturbations. Notice that on a compact space M both definitions coincide. The relation (1.1) satisfied for any bounded continuous function imposes rather strong conditions on Q_x^ϵ at infinity. Namely, for any sequence $z_n \in M$, $n = 1,2,\ldots$ such that $\inf_{i,j:i\neq j} \operatorname{dist}(z_i,z_j) > 0$ and each sequence $\epsilon_n \to 0$ one must have

$$\lim_{n\to\infty} Q_{z_n}^{\epsilon_n}(\{z_n\}) = 1$$

where $\{z\}$ is the set containing z only. Indeed, if this were not true then there exist $\rho_n > 0$, $n = 1,2,\ldots$ such that $Q_{z_n}^{\epsilon_n}(U_{\rho_n}(z_n)) < 1-\gamma$ for some $\gamma > 0$ and all $n = 1,2,\ldots$. Then acting in the same way as in the proof of (i) in Theorem 1.2 but defining a function g in neighborhoods $U_{\rho_n}(z_n)$ instead of neighborhoods of fixed radius we shall arrive to a contradiction with (1.1).

We shall define also random perturbations for continuous time dynamical systems, i.e., semiflows and flows. Let F^t, $t \geq 0$ be a one-parameter semigroup ($F^0 = \mathrm{id}$, $F^tF^s = F^{t+s}$) of continuous maps $F^t: M \to M$. When all F^t, $t \geq 0$ are invertible we consider the group, i.e., the flow, defining $F^{-t} = (F^t)^{-1}$. A family of continuous time Markov processes X_t^ϵ with transition probabilities

$$P^\epsilon(t,x,\Gamma) = P\{X_t^\epsilon \in \Gamma | X_0^\epsilon = x\}$$

will be called small random perturbations (vague small random perturbations) of a semiflow F^t if

$$\lim_{\epsilon\to 0} \sup_{x\in M} \left|\int_M P^\epsilon(t,x,dy)g(y) - g(F^t x)\right| = 0 \qquad (1.12)$$

for each $t > 0$ and any bounded $g \in C(M)$ (any $g \in C(M)$ with a compact support). A measure $\mu^\epsilon \in \mathcal{P}(M)$ will be called invariant for the process X_t^ϵ if

$$\int_M d\mu^\epsilon(x)P^\epsilon(t,x,\Gamma) = \mu^\epsilon(\Gamma) \tag{1.13}$$

for any $t > 0$ and a Borel set $\Gamma \subset M$. Similarly, a measure $\mu \in \mathcal{P}(M)$ is said to be invariant with respect to a semiflow F^t (F^t-invariant) if it is invariant with respect to all maps F^t, $t \geq 0$. Now Theorems 1.1 and 1.2 imply that all weak limits (vague limits, if M is locally compact) when $\epsilon \to 0$ of measures invariant for processes X_t^ϵ will be invariant with respect to the semiflow F^t.

One of our main goals in this book will be an attempt to specify measures which can be obtained by means of weak (vague) limits of invariant measures of Markov chains X_n^ϵ (Markov processes X_t^ϵ) when $\epsilon \to 0$. Theorems 1.1 and 1.2 (iii) say that these limiting measures must belong to the space $\mathcal{P}(M,F)$ ($\mathcal{P}(M,F^t)$) of Borel F-invariant (F^t-invariant) probability measures on M. If these spaces contain a lot of elements then the problem usually becomes rather complicated and we shall deal with it in subsequent chapters. On the other hand, if a transformation F (a semiflow F^t) is uniquely ergodic, i.e., there exists only one F-invariant (F^t-invariant) measure $\mu \in \mathcal{P}(M)$ then by Theorems 1.1 and 1.2 (iii) invariant measures of random perturbations must converge weakly (vaguely) to μ when the parameter of perturbations tends to zero. The following result yields some examples of uniquely ergodic dynamical systems.

Proposition 1.3. *Let G be a compact metrizable group.*

(i) *If $F(x) = gx$; $g,x \in G$ is a rotation on G such that the orbit $\{g^n, n = 0,\pm1,\pm2,\dots\}$ is dense in G then F is uniquely ergodic;*

(ii) *If $F^t(x) = g^t x$; $g^t, x \in G$ is a one-parameter group of rotations on G such that the orbit $\{g^t, -\infty < t < \infty\}$ is dense in G then F^t is a uniquely ergodic flow. In both cases the Haar measure is the only invariant measure.*

Proof. We shall prove both (i) and (ii) together. Let $q \in G$ then there is a sequence s_i, $i = 1,2,\ldots$ of integers in the case (i), and of reals in the case (ii) such that g^{s_i} converges to q as $i \to \infty$. Then by the dominated covergence theorem (Lebesgue theorem) for any $\mu \in \mathcal{I}(M,F)$ ($\mu \in \mathcal{I}(M,F^t)$) and each $f \in C(M)$,

$$\int f(qx)d\mu(x) = \lim_{j\to\infty} \int f(g^{s_j}x)d\mu(x)$$

$$= \lim_{j\to\infty} \int f(F^{s_j}x)d\mu(x) = \int f(x)d\mu(x).$$

This shows that μ is invariant under all rotations of G and so it is the normalized Haar measure (see Halmos [Ha], ch. 11). □

Example 1.1. Let F_α be an irrational rotation of the unit circle $\mathcal{T}^1 = \{e^{2\pi i\varphi}, 0 \leq \varphi \leq 1\}$, i.e., $F_\alpha(e^{2\pi i\varphi}) = e^{2\pi i(\varphi+\alpha)}$ with α being irrational. The orbit of this rotation is dense in $\mathcal{T}^1$, and so by Proposition 1.3, $\mathcal{I}(\mathcal{T}^1, F_\alpha)$ contains only the normalized Haar (Lebesgue in this case) measure. Now Theorems 1.1 and (1.2) say that this measure is the only limiting measure when $\epsilon \to 0$ for invariant measures of random perturbations satisfying (1.1)-(1.3).

Example 1.2. Let F_α^t, $\alpha = (\alpha_1,\ldots,\alpha_m)$ be a one-parameter group of rotations on the m-dimensional torus $\mathcal{T}^m$ defined as follows

$$F_\alpha^t(x_1,\ldots,x_m) = (x_1+\alpha_1 t(\text{mod } 1),\ldots,x_m+\alpha_m t(\text{mod } 1))$$

where $(x_1,\ldots,x_m)$ are cyclic coordinates on $\mathcal{T}^m$ with period 1 and $\alpha_1,\ldots,\alpha_m$ are rationally independent real numbers. It is easy to see that, again, each orbit of this one-parameter group of rotations is dense in $\mathcal{T}^m$. Thus the normalized Haar (= Lebesgue) measure is the only element of $\mathcal{I}(M, F_\alpha^t)$, whence it is the only limiting measure for invariant measures of random perturbations of F_α^t satisfying (1.12).

In general, neither the Markov processes X_n^ε and X_t^ε nor the dynamical systems F and F^t may have any invariant measures, at all. Nevertheless, if M is compact there is a simple sufficient condition for existence of invariant measures.

Proposition 1.4. *Let* X_n, $n = 0,1,\ldots$ *and* X_t, $t \geq 0$ *be a discrete time and a continuous time Markov processes on a compact space* M *with transition probabilities* $P(x,\Gamma) = P\{X_1 \in \Gamma | X_0 = x\}$ *and* $P(t,x,\Gamma) = P\{X_t \in \Gamma | X_0 = x\}$, *respectively. Suppose that the measures* $P(x,\cdot) \in \mathcal{P}(M)$ *and* $P(t,x,\cdot) \in \mathcal{P}(M)$ *depend on* x *continuously in the topology of weak convergence in* $\mathcal{P}(M)$. *In the continuous time case we assume also that*

$$\lim_{t\to 0} \sup_x \left| \int P(t,x,dy)g(y) - g(x) \right| = 0. \tag{1.14}$$

Then the Markov processes X_n *and* X_t *have at least one invariant measure in the sense of (1.3) and (1.13).*

Proof. Define the operators

$$Pg(x) = \int_M P(x,dy)g(y)$$

and

$$P^*\mu(\Gamma) = \int_M d\mu(x)P(x,\Gamma) \tag{1.15}$$

acting on functions and on measures on M, respectively. Remark that $(P^k)^* = (P^*)^k$. Our assumptions imply that P maps $C(M)$ into $C(M)$. Consider an arbitrary measure $\eta \in \mathcal{P}(M)$ and take $\eta_n = \frac{1}{n} \sum_{k=0}^{n-1} (P^*)^k \eta$. Since M is compact then the space $\mathcal{P}(M)$ is also compact (see Rosenblatt [Ro], p.100) and so the sequence η_n has converging subsequences. But if $\eta_{n_i} \xrightarrow{w} \rho$ then

$$P^*\eta_{n_i} = \frac{1}{n_i}\sum_{k=1}^{n_i}(P^*)^k\eta \xrightarrow{w} \rho .$$

On the other hand, for any $g \in C(M)$,

$$\int gdP^*\eta_{n_i} = \int Pgd\eta_{n_i} \xrightarrow[i\to\infty]{} \int Pgd\rho = \int gdP^*\rho$$

since $Pg \in C(M)$. Thus $P^*\eta_{n_i} \xrightarrow{w} P^*\rho$, and so $P^*\rho = \rho$, i.e., ρ is an invariant measure of the process X_n.

To prove the second part of the lemma define the operators

$$P^t f(x) = \int_M P(t,x,dy)f(y)$$

and $\qquad\qquad\qquad\qquad\qquad\qquad\qquad\qquad\qquad\qquad\qquad\qquad\qquad$ (1.16)

$$(P^t)^*\mu(\Gamma) = \int_M d\mu(x)P(t,x,\Gamma).$$

The first part of the lemma implies that for any n there exists $\rho_n \in \mathcal{P}(M)$ satisfying $(P^{\frac{1}{n}})^*\rho_n = \rho_n$. Take a subsequence $n_i \to \infty$ such that ρ_{n_i} weakly converges to some probability measure ρ. If $g \in C(M)$ then

$$\int gd(P^t)^*\rho = \int P^t gd\rho = \lim_{i\to\infty}\int P^t gd\rho_{n_i}$$

$$= \lim_{i\to\infty}\int P^{t-[tn_i]\frac{1}{n_i}} gdP^{[tn_i]\frac{1}{n_i}}\rho_{n_i}$$

$$= \lim_{i\to\infty}\int P^{t-[tn_i]\frac{1}{n_i}} gd\rho_{n_i} = \int gd\rho$$

since $\rho_{n_i} = \left[P^{\frac{1}{n_i}}\right]^* \rho_{n_i}$ and $\sup_x |P_s g(x) - g(x)| \to 0$ as $s \to 0$. □

Remark 1.2. A measure $\mu \in \mathcal{P}(M)$ is invariant for a Markov chain X_n (Markov process X_t) with transition probabilities $P(x,\Gamma)$ $(P(t,x,\Gamma))$ if and only if $P^*\mu = \mu ((P^t)^*\mu = \mu)$. By this reason we shall call such measures P^*-invariant ($(P^t)^*$-invariant). The corresponding operators for Markov chains X_n^ϵ (Markov processes X_t^ϵ) will be denoted by P_ϵ (P_ϵ^t) and so the measures μ^ϵ satisfying $P_\epsilon^*\mu^\epsilon = \mu^\epsilon$ $((P_\epsilon^t)^*\mu^\epsilon = \mu^\epsilon)$ will be called P_ϵ^*-invariant ($(P_\epsilon^t)^*$-invariant).

Corollary 1.1. *Any continuous transformation* F *and a continuous in* t *semiflow* F^t *of continuous transformations on a compact space* M *have an invariant measure.*

Proof. Define $P(x,\cdot) = \delta_{Fx}$ and $P(t,x,\cdot) = \delta_{F^t x}$ where δ_y denotes a unit mass at a point y. Now Corollary 1.1 follows from Proposition 1.4. □

Sometimes we shall be interested in invariant measures for processes on non-compact spaces M. In this case $\mathcal{P}(M)$ is not compact, and so the main ingredient in the proof of Proposition 1.4 does not work. We shall describe now a result which serves as a substitution for the compactness of $\mathcal{P}(M)$. A family $\Pi \subset \mathcal{P}(M)$ is said to be tight if for every positive ϵ there exists a compact set K such that $\mu(K) > 1-\epsilon$ for all μ in Π. One calls a family $\Pi \subset \mathcal{P}(M)$ relatively compact if every sequence of elements of Π contains a weakly convergent subsequence.

Proposition 1.5. (The Prohorov theorem) *If* Π *is tight, then it is relatively compact. If* M *is separable and complete, and* Π *is relatively compact, then* Π *is tight.*

For a proof of this result we refer the reader to Section 6 of Billingsley [Bi]. Using the above criteria one can give sufficient conditions for existence of invariant measures of Markov chains in metric spaces.

Proposition 1.6. *Let* X_n *be a Markov chain in a metric space* M *with transition probabilities* $P(x,\Gamma) = P\{X_1 \in \Gamma | X_0 = x\}$ *such that the measures* $P(x,\cdot) \in \mathcal{P}(M)$ *depend on* x *continuously in the topology of weak convergence on* $\mathcal{P}(M)$, *i.e., the operator* P^* *defined in (1.15) transforms bounded continuous functions into bounded continuous functions. Suppose that for some sequence of measures* $\sigma_n \in \mathcal{P}(M)$, $n = 1,2,\ldots$ *the family*

$\Pi = \{\mu_n : \mu_n = \frac{1}{n} \sum_{k=0}^{n-1} (P^*)^k \sigma_n,\ n = 1,2,\ldots\}$ *is tight. Then there exist* P^**-invariant probability measures.*

Proof. By Proposition 1.5 there is $\mu \in \mathcal{P}(M)$ such that $\mu_{n_j} \xrightarrow{w} \mu$ for some subsequence $n_j \to \infty$. Then for any bounded $g \in C(M)$,

$$|\int_M g dP^*\mu - \int_M g d\mu| = |\int_M Pg d\mu - \int_M g d\mu|$$

$$= \lim_{j\to\infty} |\int_M Pg d\mu_{n_j} - \int_M g d\mu_{n_j}|$$

$$= \lim_{j\to\infty} |\frac{1}{n_j} \int_M \sum_{k=0}^{n_j-1} (P^{k+1}g - P^k g) d\sigma_{n_j}|$$

$$= \lim_{j\to\infty} |\frac{1}{n_j} \int_M (P^{n_j}g - g) d\sigma_{n_j}|$$

$$\le \lim_{j\to\infty} \frac{2}{n_j} \sup_x |f(x)| = 0,$$

and so μ is P^*-invariant. □

Remark 1.3. Clearly, the family

$$\Pi = \{\mu_n = \frac{1}{n}\sum_{k=0}^{n-1} (P^*)^k\sigma_n, n = 1,2,\ldots\} \text{ is tight if}$$

$$\tilde{\Pi} = \{(P^*)^k\sigma_n; k,n = 1,2,\ldots\} \text{ is tight.}$$

The following result is well suited to random perturbations.

Theorem 1.7. *Let X_n be a Markov chain in a metric space M with transition probabilities $P(x,\Gamma) = P\{X_1 \in \Gamma | X_0 = x\}$ such that the measures $P(x,\cdot) \in \mathcal{P}(M)$ depend continuously on x in the topology of weak convergence in $\mathcal{P}(M)$. Let $F:M \to M$ be a continuous map and suppose that there exist an upper semicontinuous non-negative function $V < \infty$ on M, an integer $k > 0$ and numbers $d,\delta,R > 0$ such that the sets $K_C = \{x:V(x) \le C\}$ are compact for all numbers $C \ge 0$,*

$$V(F^kx) \le \max(R,V(x)-d) \quad \text{for any } x \in M, \tag{1.17}$$

and

$$\sup_{x\in M} \int_M P(k,x,dy)(V(y) - V(F^kx)) \le (d-\delta) \tag{1.18}$$

where $P(k,x,\Gamma) = P\{X_k \in \Gamma | X_0 = x\}$ is the k-step transition probability of the Markov chain X_n, $P(1,x,\Gamma) = P(x,\Gamma)$, and we assume that all Lebesgue integrals in (1.18) exist. Then there exists a P^*-invariant probability measure.

Proof. Remark that if the transition probabilities have the form $P(x,\Gamma) = Q_{Fx}(\Gamma)$ then using the Chapman-Kolmogorov formula

$$P(\ell+m,x,\Gamma) = \int_M P(\ell,x,dy)P(m,y,\Gamma) \tag{1.19}$$

we can derive (1.18) from the following more handy condition

$$\max_{0\le\ell\le k} \sup_{x\in M} \int_M (V(F^{\ell}y) - V(F^{\ell}x))Q_x(dy) \le (d-\delta)k^{-1}. \tag{1.20}$$

Indeed,

$$\sup_{x\in M} \int_M P(k,x,dy)(V(y) - V(F^k x)) \tag{1.21}$$

$$= \sup_{v_0\in M} \int_M \cdots \int_M Q_{Fy}(dv_1)Q_{Fv_1}(dv_2)$$

$$\times\cdots\times Q_{Fv_{k-1}}(dv_k) \sum_{i=1}^{k} (V(Fv_i^{\ell-i}) - V(F^{\ell-i+1}v_{i-1}))$$

$$\le \sum_{i=1}^{k} \sup_{v_{i-1}\in M} \int_M (V(F^{\ell-i}v_i) - V(F^{\ell-i}Fv_{i-1}))Q_{Fv_{i-1}}(dv_i)$$

$$\le d-\delta$$

where we employed (1.20) in the last step.

Next, we pass to the actual proof of Theorem 1.7. The relations (1.17) and (1.18) give that for any $y \in M\backslash K_{R+d} = \{x: V(x) > R+d\}$,

$$\int_M P(k,y,dz)V(z) \tag{1.22}$$

$$\le \int_M P(k,y,dz)V(F^k y) + \int_M P(k,y,dz)(V(z)-V(F^k y))$$

$$\le V(y)-d+d-\delta = V(y)-\delta.$$

On the other hand, if $y \in K_{R+d}$ then by (1.18),

$$\int_M P(k,y,dz)V(z) \leq \int_M P(k,y,dz)(V(z)-V(F^k y))+D \tag{1.23}$$

$$\leq d-\delta+D,$$

where $D = \sup_{y\in K_{R+d}} V(F^k y)$ which is finite since V is an upper semicontinuous function, F is a continuous map, and K_{R+d} compact.

Put

$$I(n,x) = \int_{M\setminus K_{R+d}} \cdots \int_{M\setminus K_{R+d}} P(k,x,dz_1)P(k,z_1,dz_2)$$

$$\times\cdots\times P(k,z_{n-1},dz_n)V(z_n)$$

then by (1.22) it follows that for $n > 1$,

$$I(n,x) \leq I(n-1,x) - \delta(P^k \chi_{M\setminus K_{R+d}})^{n-1} 1(x) \tag{1.24}$$

where χ_A is the indicator function of a set A, i.e., $\chi_A(x) = 1$ if $x \in A$ and $\chi_A(x) = 0$ if $x \notin A$; $1(x)$ is the function identically equal to the number 1, and for any function f the operator $(P^\ell f)$ is defined by $(P^\ell f)g = P^\ell(fg)$ with P^ℓ being the ℓ-th iteration of the operator P defined in (1.15). In these notations

$$(P^k \chi_{M\setminus K_{R+d}})^n 1(x) \tag{1.25}$$

$$= \int_{M\setminus K_{R+d}} \cdots \int_{M\setminus K_{R+d}} P(k,z_0,dz_1)\cdots P(k,z_{n-1},dz_n)$$

putting $z_0 = x$. Since $I(n,x) \geq 0$ for all $n > 1$ it follows from (1.22)-(1.24) that

$$\sum_{n=1}^{\infty} (P^k \chi_{M\setminus K_{R+d}})^n 1(x) \leq \begin{cases} (d+D)\delta^{-1} & \text{if } x \in K_{R+d} \\ \delta^{-1} V(x) & \text{if } x \notin K_{R+d} \end{cases}. \tag{1.26}$$

Next, we shall follow the proof of the Foguel theorem [Fo] (see also Rosenblatt [Ro], p.102) that enables one to construct a σ-finite P^*-invariant measure which in our case will be finite. Notice that (1.26) implies

$$(P\chi_{M\setminus K_{R+d}})^n 1(x) \leq (P^k\chi_{M\setminus K_{R+d}})^{[\frac{n}{k}]} 1(x) \to 0 \qquad (1.27)$$

$$\text{as} \quad n \to \infty,$$

where $[a]$ denotes the integral part of a number a. If $V(y) \leq R+d$ for all $y \in M$ then $K_{R+d} = M$ and according to Proposition 1.4 there is nothing to prove. Suppose that $V(z) > R+d$ for some $z \in M$ then for some $\gamma > 0$ the set $Q = \{y : V(y) \geq R+d+\gamma\}$ is non-empty and closed (since V is upper semicontinuous). By the Urysohn lemma there exists a continuous function $q(y)$ such that $0 \leq q \leq 1$ and $q(y) = 1$ if $y \in K_{R+d}$ and $q(y) = 0$ if $y \in M\setminus K_{R+d+\gamma}$. Set $r = 1-q$ and introduce the operator

$$P_N = \sum_{n=0}^{N} (Pr)^n Pq$$

acting on bounded measurable functions on $K_{R+d+\gamma}$. Clearly, P_N is a positive operator, i.e., $P_N f \geq 0$ if $f \geq 0$, and P_N preserves the space of bounded continuous functions on $K_{R+d+\gamma}$. Furthermore, the sequence P_N is nondecreasing in the sense that $P_{N+1}f \geq P_N f$ if $f \geq 0$, and

$$P_N\chi_{K_{R+d+\gamma}} = \sum_{n=0}^{N} (Pr)^n Pq = \sum_{n=0}^{N} (Pr)^n(1-Pr) \qquad (1.28)$$

$$= P1-(Pr)^{N+1} \leq 1.$$

Next, by (1.27),

$$(Pr)^N 1 \leq (P\chi_{M\setminus K_{R+d}})^N 1 \to 0 \quad \text{as} \quad N \to \infty, \tag{1.29}$$

and so $(Pr)^N 1$ converges monotonously to zero on $K_{R+d+\gamma}$ since $(Pr)1 \leq 1$. The convergence in (1.29) must therefore be uniform on $K_{R+d+\gamma}$ since this set is compact. Furthermore, similarly to (1.28)

$$\lim_{N\to\infty} P_N 1 = 1 - \lim_{N\to\infty}(Pr)^{N+1} 1 = 1$$

and the convergence is uniform on $K_{R+d+\gamma}$. Thus

$$\|P_{N+m} f - P_N f\| = \|\sum_{n=N+1}^{N+m} (Pr)^n (Pq) f\|$$

$$\leq \|f\| \cdot \|P_{N+m} 1 - P_N 1\| \to 0 \quad \text{as} \quad N \to \infty$$

where $\|g\| = \sup_y |g(y)|$. Thus there exist a uniform limit

$$P_\infty = \lim_{N\to\infty} P_N$$

which is a positive operator mapping $P_\infty : C(K_{R+d+\gamma}) \to C(K_{R+d+\gamma})$ and satisfying $P_\infty 1 = 1$. By a version of Proposition 1.4 there exists a P_∞^*-invariant probability measure λ on $K_{R+d+\gamma}$, i.e., $P_\infty^* \lambda = \lambda$ where $\int f dP_\infty^* \lambda = \int P_\infty f d\lambda$ for any bounded Borel function f on $K_{R+d+\gamma}$.

Define,

$$\nu = \sum_{n=0}^{\infty} ((Pr)^*)^n \lambda \tag{1.30}$$

where, again, $(Pr)^*\lambda$ is defined by $\int fd(Pr)^*\lambda = \int (Pr)fd\lambda$ which holds true for every bounded Borel function f. Notice that if f is zero outside K_{R+d} then $(Pr)^n f = 0$ for $n = 1,2,\dots$, and so $\int fd\nu = \int fd\lambda$. In particular, taking $f = \chi_{K_{R+d}}$ we obtain $\nu(K_{R+d}) = \lambda(K_{R+d}) \leq 1$. On the other hand, using (1.26) we derive from (1.30) that

$$\nu(M\backslash K_{R+d}) = \int_{R+d+\gamma} \sum_{n=0}^{\infty} (Pr)^n \chi_{M\backslash K_{R+d}} d\lambda$$

$$\leq \int_{K_{R+d+\gamma}} \sum_{n=0}^{\infty} (P\chi_{M\backslash K_{R+d}})^n d\lambda$$

$$\leq k(1 + \int_{K_{R+d+\gamma}} \sum_{n=1}^{\infty} (P^k \chi_{M\backslash K_{R+d}})^n d\lambda)$$

$$\leq k(1 + (d+D)\delta^{-1} + \delta^{-1} \sup_{x\in K_{R+d+\gamma}} V(x)) < \infty$$

which is finite since V is upper semicontinuous and $K_{R+d+\gamma}$ is compact. Hence ν is a finite measure, and so $\mu = \frac{1}{\nu(M)} \nu \in \mathcal{P}(M)$. It suffices to show that ν is P^*-invariant.

Indeed, by (1.30)

$$P^*\nu = \sum_{n=0}^{\infty} P^*((Pr)^*)^n\lambda$$

$$= \sum_{n=0}^{\infty} ((Pr)^*)^{n+1}\lambda + \sum_{n=0}^{\infty} (Pq)^*((Pr)^*)^n\lambda$$

$$= \nu - \lambda + P_\infty^*\lambda = \nu - \lambda + \lambda = \nu$$

since $P_\infty^* = \sum_{n=0}^{\infty} ((Pr)^n(Pq))^* = \sum_{n=0}^{\infty} (Pq)^*((Pr)^*)$ and $P_\infty^*\lambda = \lambda$. This completes the proof of Theorem 1.7. □

Remark 1.4. The condition (1.17) means that the compact set $K_R = \{x:V(x) \le R\}$ is invariant under F^k and the whole space M is being attracted towards K_R. If V is continuous then the condition (1.20) will be satisfied if $P(x,\cdot) = Q_{Fx}(\cdot)$ and the measures Q_x are close to unit masses concentrated at x, and so this assumption is suited well to random perturbations. In a metric space one of natural candidates for the function $V(x)$ is $\mathrm{dist}(x_0,x)$ where x_0 is a fixed point.

Corollary 1.2. *Let a family of Markov chains* X_n^ε *be vague random perturbations of a continuous map* $F:M \to M$ *with transition probabilities* $P^\varepsilon(x,\cdot) = Q_{Fx}^\varepsilon(\cdot)$ *such that* M, F *and* Q_y^ε, $\varepsilon > 0$ *all satisfy conditions of Theorem 1.7. Then all vague limit points as* $\varepsilon \to 0$ *of invariant measures of* X_n^ε *have support in* K_R.

Proof. Since all sets $K_C = \{x:V(x) \le C\}$ are compact, the sets $G_C = \{v:V(x) < C\}$ are open, and $M = \bigcup_{C \ge 0} G_C$ then M is locally compact. By Theorem 1.2 (iii) all vague limits of $(P^\varepsilon)^*$-invariant measures (which exist by Theorem 1.7) are F-invariant. By (1.17), $F^{-nk}K_R \supset K_{R+nd}$ which together with (1.5) mean that if μ is F-invariant then $\mu(K_R) = \mu(K_{R+nd})$ for all $n = 1,2,\dots$. Since $K_{R+nd} \uparrow M$ as $n \uparrow \infty$ we obtain $\mu(K_R) = \mu(M) = 1$ proving Corollary 1.2. □

There is another assumption called Doeblin's condition which does not make use of topology and ensures existence and uniqueness of invariant measures for Markov chains. If random perturbations X_n^{ϵ} satisfy Doeblin's condition then for each ϵ there is only one invariant measure μ^{ϵ} of X_n^{ϵ}, and so the limiting behavior of μ^{ϵ} as $\epsilon \to 0$ becomes even more interesting since, usually, there are a lot of F-invariant measures which enforce the question about the right candidate for a limit of measures μ^{ϵ}.

Proposition 1.8. (cf. Doob [Do], p.197) *Let there exist a measure* $\upsilon \in \mathcal{P}(M)$, *an integer* $k > 0$, *a number* $\delta > 0$, *and a Borel set* G *(in this result the topology does not play any role and so a Borel set means that this set belongs to a fixed σ-field whose members are called measurable sets) such that*

$$\upsilon(G) > 0 \quad \text{and} \quad p_0(k,x,y) \geq \delta \quad \text{for any} \quad x \in M \quad \underline{\text{and}}\ y \in G \tag{1.31}$$

where $p_0(k,x,\cdot)$ *is the density of the absolutely continuous component of* $P(k,x,\cdot) \in \mathcal{P}(M)$ *with respect to* υ, *and, again,* $P(\ell,y,\Gamma)$ *denotes the transition probability of a Markov chain* X_n *on* M. *Then there is a* P^**-invariant measure* $\mu \in \mathcal{P}(M)$ *such that* $\mu(\tilde{G}) \geq \delta\upsilon(\tilde{G})$ *for any Borel* $\tilde{G} \subset G$, *and*

$$|P(n,x,Q) - \mu(Q)| \leq (1-\delta\upsilon(G))^{(n/k)-1}, \quad n = 1,2,\ldots \tag{1.32}$$

for each Borel $Q \subset M$, *and so such* μ *is unique.*

Proof. For any Borel set $Q \subset M$ define

$$b_Q^{(n)} = \inf_x P(n,x,Q) \quad \text{and} \quad B_Q^{(n)} = \sup_x P(n,x,Q). \tag{1.33}$$

By the Chapman-Kolmogorov formula (1.19),

$$B_Q^{(n+1)} = \sup_x \int_M P(x,dy)P(n,y,Q)$$

$$\le \sup_x \int_M P(x,dy)B_Q^{(n)} = B_Q^{(n)}.$$

Similarly, $b_Q^{(n+1)} \ge b_Q^{(n)}$. Therefore the following limits

$$B_Q = \lim_{n\to\infty} B_Q^{(n)} \ge b_Q = \lim_{n\to\infty} b_Q^{(n)}$$

exist. For fixed $x,y \in M$ introduce the set function

$$\Psi(Q) = P(k,x,Q) - P(k,y,Q).$$

By the Hahn decomposition theorem (see, for instance, Halmos [Ha]) there exist Borel sets S_+ and $S_- = M \setminus S_+$ such that $\Psi(V) \ge 0$ and $\Psi(\tilde{V}) \le 0$ for any Borel $V \subset S_+$ and $\tilde{V} \subset S_-$. Since $P(k,x,M) = P(k,y,M) = 1$, then

$$\Psi(S_+) + \Psi(S_-) = \Psi(M) = 0 \tag{1.34}$$

and

$$\Psi(S_+) = P(k,x,S_+) - P(k,y,S_+) \tag{1.35}$$

$$= 1 - P(k,x,S_-) - P(k,y,S_+)$$

$$\le 1 - \int_{S_-} p_0(k,x,z)d\upsilon(z) - \int_{S_+} p_0(k,y,z)d\upsilon(z)$$

$$\le 1 - \delta\upsilon(G),$$

where we used (1.31) and the fact that taking only absolutely continuous components of $P(k,x,\cdot)$ and $P(k,y,\cdot)$ with respect to υ wc subtract less. Now (1.34) and (1.35) imply that for any Borel Q,

$$P(n+k,x,Q) - P(n+k,y,Q) \tag{1.36}$$

$$= \int_M (P(k,x,dz) - P(k,y,dz))P(n,z,Q)$$

$$\le B_Q^{(n)} \int_{S_+} \Psi(dz) + b_Q^{(n)} \int_{S_-} \Psi(dz)$$

$$= \Psi(S_+)(B_Q^{(n)} - b_Q^{(n)}) \le (1 - \delta\nu(G))(B_Q^{(n)} - b_Q^{(n)}).$$

Since $x,y \in M$ are arbitrary points, (1.36) means that

$$B_Q^{(n+k)} - b_Q^{(n+k)} \le (1 - \delta\nu(G))(B_Q^{(n)} - b_Q^{(n)}),$$

and so

$$B_Q^{(\ell k)} - b_Q^{(\ell k)} \le (1 - \delta\nu(G))^{\ell}.$$

Therefore $B_Q^{(n)}$ and $b_Q^{(n)}$ must have a common limit $\mu(Q)$ as $n \to \infty$ and

$$|P(n,x,Q) - \mu(Q)| \le B_Q^{(n)} - b_Q^{(n)} \le (1 - \delta\nu(G))^{(n/k)-1}.$$

If $\tilde{G} \subset G$ then $\mu(\tilde{G}) \ge b_{\tilde{G}}^{(k)} \ge \delta\nu(\tilde{G})$ proving Proposition 1.8. □

Remark 1.5. The above result is efficient when transition probabilities have densities with respect to a fixed common measure. This will be the case for some types of random perturbations, which we shall consider in subsequent chapters, in particular, for diffusion type perturbations. Other useful conditions for existence of invariant measures of diffusion processes via their generators can be found in Hasminskii [Has].

1.2. Entropy via random perturbations: generalities.

In this section we shall see how random perturbations can help to estimate the entropies of dynamical systems.

First, we shall review basic definitions and properties of the measure theoretic (metric) entropy of a measure preserving transformation which we may need in the subsequent exposition. We shall only formulate the main results referring the reader to Walters [Wa] for proofs and further details. We shall use facts concerning entropy in Section 2.5. The reader who is not interested in ergodic theory may pass directly to the next section.

Let M be a space with a given σ-field $\mathcal{B}$ of measurable sets and let μ be a probability measure on M. A disjoint collection of elements of $\mathcal{B}$ whose union is M will be called a partition of M. The join of two partitions $\xi = \{A_1,\ldots,A_n\}$ and $\eta = \{C_1,\ldots,C_k\}$ is the partition $\xi \vee \eta = \{A_i \cap C_j : 1 \le i \le n, 1 \le j \le k\}$. If $\varphi : M \to M$ is a measurable map and $\xi = \{A_1,\ldots,A_k\}$ is a partition then $\varphi^{-1}\xi$ denotes the partition $\{\varphi^{-1}A_1,\ldots,\varphi^{-1}A_k\}$. The entropy of a partition $\xi = \{A_1,\ldots,A_k\}$ is defined as the number

$$H_\mu(\xi) = -\sum_{i=1}^{k} \mu(A_i)\log \mu(A_i). \tag{2.1}$$

Here and throughout this book $\log a$ means the natural logarithm of a and the expression $0 \log 0$ is considered to be 0. Let a measurable map $\varphi : M \to M$ preserves the measure μ in the sense that $\mu(\varphi^{-1}\Gamma) = \mu(\Gamma)$ for any $\Gamma \in \mathcal{B}$. The entropy of φ with respect to a partition $\xi = \{A_1,\ldots,A_k\}$ is defined as the limit

$$h_\mu(\varphi,\xi) = \lim_{n\to\infty} \frac{1}{n} H_\mu\Big(\bigvee_{i=0}^{n-1} \varphi^{-i}\xi\Big) \tag{2.2}$$

$$= \inf_{n\geq 0} \frac{1}{n} H_\mu\Big(\bigvee_{i=0}^{n-1} \varphi^{-i}\xi\Big)$$

which exists since $c_n = H_\mu(\bigvee_{i=0}^{n-1} \varphi^{-i}\xi)$ turns out to be a subadditive sequence (see Walters [Wa], p.88). Finally, the entropy of a map $\varphi:M \to M$ preserving a measure $\mu \in \mathcal{P}(M)$ is defined by

$$h_\mu(\varphi) = \sup_\xi h_\mu(\varphi,\xi) \tag{2.3}$$

where the supremum is taken over all finite partitions of M. We remark, at once, that all definitions remain unchanged if we consider also countable partitions. It is easy to see that this will lead to the same entropy.

The calculation of entropy can be simplified if one uses the following Kolmogorov-Sinai theorem (see Walters [Wa], 95-96).

Proposition 2.1. *Let* $\varphi:M \to M$ *preserves a measure* $\mu \in \mathcal{P}(M)$ *and let* ξ *be a finite partition such that the minimal* σ*-field containing* $\bigvee_{i=0}^{\infty} \varphi^{-i}\xi$ *coincides with the* σ*-field of measurable sets* $\mathcal{B}$. Then $h_\mu(\varphi) = h_\mu(\varphi,\xi)$. *If* φ *is invertible then* $h_\mu(\varphi) = h_\mu(\varphi,\xi)$ *if the minimal* σ*-field containing* $\bigvee_{i=-\infty}^{\infty} \varphi^{i}\xi$ *coincides with* $\mathcal{B}$.

The partitions satisfying the conditions of Proposition 2.1 are called generators. For one class of measure preserving transformations the construction of generators is especially easy. A homeomorphism φ of a compact metric space M is said to be expansive if there exists $\delta > 0$ called an expansive constant for φ such that if $x \neq y$ then $\mathrm{dist}(\varphi^n x,\varphi^n y) > \delta$ for some integer n.

Proposition 2.2. (see Walters [Wa], p.143) *Let* φ *be an expansive homeomorphism of a compact metric space* M *with an expansive constant* δ. *If* $\xi = \{A_1, \dots, A_k\}$ *is a partition of* M *into Borel sets with* $\operatorname{diam}(A_j) \le \delta$, $1 \le j \le k$, *then the minimal* σ*-field containing* $\bigvee_{i=-\infty}^{\infty} \varphi^i \xi$ *coincides with the Borel* σ*-field* $\mathcal{B}(M)$ *on* M. *Thus if* φ *preserves a measure* $\mu \in \mathcal{P}(M)$ *then* $h_\mu(\varphi) = h_\mu(\varphi, \xi)$.

The following version of the Shannon-McMillan-Breiman theorem is often useful for various estimates concerning the entropy.

Proposition 2.3. *Let* $\varphi: M \to M$ *be a measurable map preserving a measure* $\mu \in \mathcal{P}(M)$. *For a finite partition* ξ *denote by* $\xi_n(x)$ *the element of the partition* $\bigvee_{i=0}^{n-1} \varphi^{-i} \xi$ *which contains* $x \in M$. *Then for* μ*-almost all* $x \in M$ *the limit* $r(x) = \lim_{n\to\infty} \frac{1}{n} \log \mu(\xi_n(x))$ *exists and* $h_\mu(F, \xi) = -\int_M r(x) d\mu(x)$.

For the proof we refer the reader to Parry [Pa].

Next, we shall go back to Markov chains X_n^ϵ which are small random perturbations of a transformation $F: M \to M$ in the sense of (1.1)-(1.2). Recall the construction of the probability R^ϵ on the sample space Ω corresponding to the Markov chain X_n^ϵ having an invariant measure μ^ϵ. First, R^ϵ is defined on the sets of the form

$$\pi(G_0, \dots, G_k) = \{\omega: X_0(\omega) \in G_0, \dots, X_k(\omega) \in G_k\} \tag{2.4}$$

by

$$R^\epsilon(\pi(G_0, \dots, G_k)) \tag{2.5}$$

$$= \int_{G_0} \dots \int_{G_k} d\mu^\epsilon(x) P^\epsilon(x, dx_1) \dots P^\epsilon(x_{k-1}, dx_k)$$

for any Borel sets $G_i \subset M$. The sample space Ω can be identified with the infinite product $M^{\mathbb{N}} = M \times \ldots \times M \times \ldots$ and employing Ionescu-Tulcea's or Kolmogorov's extension theorems (see Neveu [Ne]) one obtains R^ϵ defined already on all measurable subsets of $M^{\mathbb{N}} = \Omega$. Define $\theta:\Omega \to \Omega$ by $X_n(\theta\omega) = X_{n+1}(\omega)$, $n = 0,1,\ldots$, then $R^\epsilon \in \mathcal{P}(\Omega)$ is a θ-invariant measure since we assume that μ^ϵ is an invariant measure of ξ_n^ϵ. Indeed, $\theta^{-1}\pi(G_0,\ldots,G_k) = \{\omega:X_0(\theta\omega) \in G_0, X_1(\theta\omega) \in G_1,\ldots, X_k(\theta\omega) \in G_k\} = \{\omega:X_0(\omega) \in M, X_1(\omega) \in G_0,\ldots,X_{k+1}(\omega) \in G_k\}$ and so

$$R^\epsilon(\theta^{-1}\pi(G_0,\ldots,G_k)) \tag{2.6}$$

$$= \int_M d\mu^\epsilon(x)P^\epsilon(x,dx_1)\int_{G_0}\cdots\int_{G_k} P^\epsilon(x_1,dx_2)\ldots P^\epsilon(x_k,dx_{k+1})$$

$$= \int_{G_0}\cdots\int_{G_k} d\mu^\epsilon(x_1)P^\epsilon(x_1,dx_2)\ldots P^\epsilon(x_k,dx_{k+1})$$

$$= R^\epsilon(\pi(G_0,\ldots,G_k)).$$

This equality being true for all $\pi(G_0,\ldots,G_k)$ implies already that R^ϵ is θ-invariant.

Considering now the measure preserving transformation θ on Ω we can define entropies $h^\epsilon(\theta,\xi)$ with respect to finite (countable) partitions ξ of Ω by (2.2) with $\mu = R^\epsilon$, as well as the entropy $h^\epsilon(\theta) = \sup_\xi h^\epsilon(\theta,\xi)$.

Suppose that $\mu^{\epsilon_i} \xrightarrow{w} \mu$ along a subsequence $\epsilon_i \to 0$ then by Theorem 1.1 μ is F-invariant and so we can define the entropies $h_\mu(F,\Pi)$ with respect to finite partitions Π of M, and the entropy $h_\mu(F) = \sup_\Pi h_\mu(F,\Pi)$ of F. The natural desire is to understand the connection between the limiting behavior of $h^{\epsilon_i}(\theta,\xi)$, $h^{\epsilon_i}(\theta)$ as $\epsilon_i \to 0$, and $h_\mu(F,\Pi)$, $h_\mu(F)$. The following result shows that the "absolute" entropy $h^\epsilon(\theta)$ is not a good notion for many Markov chains and it will not be of any help for our purposes.

Theorem 2.4. *Suppose that all transition probabilities* $P^\epsilon(x,\cdot)$ *have bounded densities* $p^\epsilon(x,y) \le K < \infty$ *(*ϵ *is fixed here) with respect to some measure* $m \in \mathcal{P}(M)$, *i.e.,* $P^\epsilon(x,G) = \int_G p^\epsilon(x,y)dm(y)$ *for any Borel* $G \subset M$. *Assume that for any* $n \ge 1$ *there exists a partition* $\Pi_n = \{A_1^{(n)},\dots,A_{k_n}^{(n)}\}$ *such that* $m(A_i^{(n)}) \le \frac{1}{n}$ *for all* $i = 1,\dots,k_n$. *Then* $h^\epsilon(\theta) = \infty$.

Proof. Consider the family of partitions $\zeta_n = (\Gamma_1^{(n)},\dots,\Gamma_{k_n}^{(n)})$ of Ω such that $\Gamma_i^{(n)} = \{\omega : X_0^\epsilon(\omega) \in A_i^{(n)}\}$. Then clearly $\bigvee_{j=0}^{\ell-1} \theta^{-j}\zeta_n$ is the partition into sets $\pi(A_{i_0}^{(n)},\dots,A_{i_{\ell-1}}^{(n)})$ where $\pi(\cdot)$ was defined by (2.4) and $A_{i_j}^{(n)} \in \Pi_n$, $1 \le i_j \le k_n$. Thus by (2.5),

$$H^\epsilon\Big(\bigvee_{j=0}^{\ell-1} \theta^{-j}\zeta_n\Big) \tag{2.7}$$

$$= -\sum_{i_0,\dots,i_{\ell-1}} R^\epsilon(\pi(A_{i_0}^{(n)},\dots,A_{i_{\ell-1}}^{(n)}))\log R^\epsilon(\pi(A_{i_0}^{(n)},\dots,A_{i_{\ell-1}}^{(n)}))$$

$$\ge -\log(K \max_i m(A_i^n))^\ell \ge \ell \log \frac{n}{k}.$$

Thus $h^\epsilon(\theta) \ge h^\epsilon(\theta,\zeta_n) \ge \log \frac{n}{k}$ for any $n = 1,2,\dots$ and so $h^\epsilon(\theta) = \infty$. □

Remark 2.1. For every non-atomic Borel probability measure μ (i.e., $\mu(\{x\}) = 0$ for each singleton x) on a complete separable metric space M one can choose a sequence of partitions $\Pi_n = \{A_1^{(n)}, \dots, A_{k_n}^{(n)}\}$ such that $m(A_i^{(n)}) \le \frac{1}{n}$ for all $i = 1, \dots, k_n$. Indeed, any such measure μ is tight (see Billingsley [Bi], Theorem 1.4), i.e., for any $n > 0$ there is a compact set K such that $\mu(M \setminus K) \le \frac{1}{n}$. Since μ is non-atomic each point has an open neighborhood whose μ-measure is less than $\frac{1}{n}$. Finite number of these cover K. Taking differences and intersections we shall pass from the cover to a partition with required properties.

The existence of transition densities p^ϵ as required in Theorem 2.4 is a rather natural assumption in the theory of Markov chains. So the disappointing result of Theorem 2.4 suggests to try another quantity, namely entropies with respect to partitions.

Theorem 2.5. *Let* X_n^ϵ *be random perturbations of a continuous map* $F: M \to M$ *of a metric space* M *and let* $P_{\epsilon_i}^*$*-invariant measures* μ^{ϵ_i} *(see Remark 1.2) weakly converge to* μ *as* $\epsilon_i \to 0$. *Suppose that* $\Pi = (V_1, \dots, V_k)$ *is a partition of* M *into Borel sets such that* $\mu(\partial\Pi) = 0$ *where* $\partial\Pi = \bigcup_{i=1}^{k} \partial V_i$ *and* ∂ = *closure"interior denotes the boundary of a set. Then*

$$\limsup_{i\to\infty} h^{\epsilon_i}(\theta, \zeta^{(i)}) \le h_\mu(F, \Pi) \le h_\mu(F) \qquad (2.8)$$

where $\zeta^{(i)}$ *is the partition of* Ω *into the sets* $\Gamma_j^{(i)} = \{\omega : X_0^{\epsilon_i} \in V_j\}$. *In other words,*

$$\limsup_{i\to\infty} \lim_{n\to\infty}\Big(-\frac{1}{n} \sum_{i_0,\dots,i_{n-1}} R^{\epsilon_i}(\pi(V_{i_0},\dots,V_{i_{n-1}}))$$

$$\times \log R^{\epsilon_i}(\pi(V_{i_0},\dots,V_{i_{n-1}}))\Big) \tag{2.9}$$

$$\leq h_\mu(F,\Pi)$$

where R^ϵ *is defined by* (2.5).

Proof. Define probability measures υ_n^ϵ and υ_n on the n-fold product $M^n = M \times \dots \times M$ by

$$\upsilon_n^\epsilon(G_1 \times \dots \times G_n) = R^\epsilon(\pi(G_1,\dots,G_n)) \tag{2.10}$$

and

$$\upsilon_n(G_1 \times \dots \times G_n) = \mu\Big(\bigcap_{0\leq j\leq n-1} F^{-j}G_{j+1}\Big) \tag{2.11}$$

for any Borel sets $G_i \subset M$, $i = 1,\dots,n$ and extending these measures to the whole Borel σ-field on M^n. We claim that $\mu^{\epsilon_i} \xrightarrow{w} \mu$ as $\epsilon_i \to 0$ implies $\upsilon_n^{\epsilon_i} \xrightarrow{w} \upsilon_n$ as $\epsilon_i \to 0$ for each $n = 1,2,\dots$. Indeed, for any bounded continuous functions $g_1,\dots,g_n$ on M one has

$$\int_{M^n} g_1(z_1)\cdots g_n(z_n)\upsilon_n^{\epsilon_i}(dz_1 \times \cdots \times dz_n) \tag{2.12}$$

$$= \int_M \cdots \int_M d\mu^{\epsilon_i}(z_1)g_1(z_1)P^\epsilon(z_1,dz_2)g_2(z_2)$$

$$\times \cdots \times P^\epsilon(z_{n-1},dz_n)g_n(z_n).$$

In view of (1.1), (1.2) and $\mu^{\epsilon_i} \xrightarrow{w} \mu$ the last integral in (2.12) converges as $\epsilon_i \to 0$ to

$$\int_M d\mu(z)g_1(z)g_2(Fz)\cdots g_n(F^{n-1}z) = \int_{M^n} g_1\cdots g_n d\upsilon_n$$

which proves $\nu_n^{\epsilon_i} \xrightarrow{w} \nu_n$ as $i \to \infty$.

Since $\mu(\partial\Pi) = 0$ and μ is F-invariant then $\mu(\bigcup_{j=0}^{n-1} F^{-j}\partial\Pi) = 0$. Thus by (2.11) one has $\nu_n(\partial(V_{i_0} \times \cdots \times V_{i_{n-1}})) = 0$ for any $1 \leq i_j \leq k$ where $V_{i_j} \in \Pi$. By the basic result about the weak convergence (see Theorem 2.1 (v) in Billingsley [Bi]) it follows from here that

$$\nu_n^{\epsilon_i}(V_{i_0} \times \cdots \times V_{i_{n-1}}) = R^{\epsilon_i}(\pi(V_{i_0},\ldots,V_{i_{n-1}})$$
$$\longrightarrow \mu(\bigcap_{0\leq j\leq n-1} F^{-j}V_{i_j}) \quad \text{as} \quad \epsilon_i \to 0. \tag{2.13}$$

Denote

$$H_n^{\epsilon}(\Pi) = - \sum_{1\leq i_0,\ldots,i_{n-1}\leq k} R^{\epsilon_i}(\pi(V_{i_0},\ldots,V_{i_{n-1}})) \tag{2.14}$$
$$\times \log R^{\epsilon_i}(\pi(V_{i_0},\ldots,V_{i_{n-1}}))$$

and

$$H_n(\Pi)$$
$$= - \sum_{i_0,\ldots,i_{n-1}} \mu(\bigcap_{0\leq j\leq n-1} F^{-j}V_{i_j})\log \mu(\bigcap_{1\leq j\leq n-1} F^{-j}V_{i_j}). \tag{2.15}$$

As we have already pointed out the sequences $c_n^{\epsilon} = H_n^{\epsilon}(\Pi)$ and $c_n = H_n(\Pi)$ are subadditive (see Walters [Wa], p.88) and so

$$h^{\epsilon_i}(\theta,\zeta^{(i)}) = \lim_{n\to\infty} \frac{1}{n} H_n^{\epsilon_i}(\Pi) = \inf_{n>0} \frac{1}{n} H_n^{\epsilon_i}(\Pi) \tag{2.16}$$

and

$$h_\mu(F,\Pi) = \lim_{n\to\infty} \frac{1}{n} H_n(\Pi) = \inf_{n>0} \frac{1}{n} H_n(\Pi). \tag{2.17}$$

Now (2.13)-(2.17) yield,

$$\begin{aligned}\limsup_{i\to\infty} h^{\epsilon_i}(\theta,\zeta^{(i)}) &\le \inf_{n>0} \frac{1}{n} \limsup_{i\to 0} H_n^{\epsilon_i}(\Pi) \\ &= \inf_{n>0} \frac{1}{n} H_n(\Pi) = h_\mu(F,\Pi)\end{aligned} \tag{2.18}$$

proving Theorem 2.5. □

In the next chapter we shall see that for certain hyperbolic dynamical systems one can achieve in (2.8) an equality.

Remark 2.2. Theorem 2.5 remains true when Π is a countable partition with $\mu(\partial\Pi) = 0$. When striving for an equality in (2.8) one first has to choose a partition Π (finite or countable) such that $\mu(\partial\Pi) = 0$ and $h(F,\Pi) = h(F)$. If F is an expansive homeomorphism then by Proposition 2.2 to satisfy the last equality it suffices to choose a partition with elements of small diameter. To construct such partitions one notices that each point x has arbitrarily small balls $U_\rho(x)$ centered at x whose boundaries have μ-measure zero (since only countably many balls centered at x may have boundaries of positive μ-measure). Taking one such ball at each x we obtain an open cover of M. If M is compact then we can choose a finite subcover and if M is not compact but separable we can choose a countable subcover (by the Lindelöf theorem - see Kelley [Ke], p.49). Having such cover $\{Q_1,Q_2,...\}$ one constructs a partition $\Pi = \{V_1,V_2,...\}$ defining $V_1 = Q_1$, $V_{n+1} = Q_{n+1}\setminus(\bigcup_{i=1}^{n} V_i)$. Clearly, $\mu(\partial\Pi) = 0$ and the elements of Π can be made of arbitrarily small size.

Remark 2.3. It is often convenient to consider partitions $\Pi = \{V_1, \ldots, V_k\}$ whose elements are sets satisfying certain good properties. Then it is not always possible to get the disjointness $V_i \cap V_j = \phi$, $i \neq j$. Still, if $V_i \cap V_j \subset \partial V_i \cap \partial V_j$, $i \neq j$ and $\mu(\partial\Pi) = 0$ then all our arguments concerning the entropy, as well, as Theorem 2.5 go through. We shall use such collections of sets in the next chapter calling them also partitions.

Remark 2.4. If F^t, $-\infty < t < \infty$ is a continuous time dynamical system (i.e., a flow) preserving a measure μ then by the definition the entropy of the flow is the entropy of its time-one map F^1. This is justified by the formula

$$h_\mu(F^t) = |t| h_\mu(F^1) \tag{2.19}$$

for any t (see Walters [Wa], p.91 for integer t and Kifer [Ki9], p.177 for any t).

1.3. Locating invariant sets.

In this section we shall discuss the connection between certain parameters of random perturbations and invariant sets of dynamical systems. We shall continue this study in Sections 2.7, 3.1, and 3.2. Our exposition here follows the author's paper [Ki8].

We shall consider simultaneously both random perturbations X_n^ϵ of a homeomorphism $F: M \to M$ and continuous time random perturbations X_t^ϵ of continuous in t group of homeomorphisms F^t (the flow). In both cases we shall write X_t^ϵ and F^t assuming that $t \in T$ with T being either $[0,\infty)$ or $\mathbb{N} = \{0,1,2,\ldots\}$. Let $K \subset M$ be a compact subset of a complete locally compact metric space M. A set $\Lambda(K) \subset K$ is called the maximal invariant (under the action of F^t) set in K if any set G satisfying the property

$$F^t G = G \subset K \quad \text{for all} \quad t \in (-\infty, \infty) \tag{3.1}$$

is a subset of $\Lambda(K)$. Obviously, $\Lambda(K)$ is a compact set (maybe empty) since F^t, $t \in T$ are continuous and K is compact. Denote by τ the exit time from K for the process X_t^ϵ, i.e.,

$$\tau = \inf\{t \in T : X_t^\epsilon \notin K\}. \tag{3.2}$$

We shall prove the following result. As usual, we denote by $P_x^\epsilon\{A\}$ and $E_x^\epsilon \xi$ the probability of the event A and the expectation of the random value ξ, respectively, for the process X_t^ϵ starting at x.

Theorem 3.1. (a) *If for some* $x \in \text{int}\, K$ (int *means interior*),

$$\limsup_{\epsilon \to 0} \limsup_{t \to \infty} \frac{1}{t} \log P_x^\epsilon\{\tau > t\} > -\infty \tag{3.3}$$

then the maximal invariant set $\Lambda(K)$ *is not empty;*

(b) *If for some* $x \in \text{int}\, K$,

$$\limsup_{\epsilon \to 0} E_x^\epsilon \tau = \infty, \tag{3.4}$$

then the set $\Lambda(K)$ *is not empty.*

Proof. Suppose $\Lambda(K)$ is empty. Then there exists a compact set $\tilde{K}$ such that $\text{int}\, \tilde{K} \supset K$ and $\Lambda(\tilde{K})$ is also empty. Indeed, since we assume M to be locally compact each point $x \in K$ has an open ball $U^{(n)}(x)$ of radius less than $\frac{1}{n}$ centered at x whose closure $\overline{U^{(n)}}(x)$ is compact. Open balls $U^{(n)}(x)$ cover K and so one can choose a finite open subcover $U^{(n)}(x_i^{(n)})$, $i = 1,\ldots,\ell_n$. Now consider the sequence of compact sets

$$K_1 = \bigcup_{1\leq i\leq \ell} \overline{U^{(1)}(x_i^{(1)})},\ K_n = K_{n-1} \cap \Big(\bigcup_{1\leq i\leq \ell_n} \overline{U^{(n)}(x_i^{(n)})}\Big)$$

for $n = 2,3,\ldots$. Then, clearly,

$$K_1 \supset K_2 \supset \ldots \quad \text{and} \quad \bigcap_{m\geq 1} K_m = K. \tag{3.5}$$

Furthermore, $\Lambda(K_1) \supset \Lambda(K_2) \supset \ldots \supset \Lambda(K)$. If $\Lambda(K_m) \neq \phi$ for all $m \geq 1$ then

$$\Lambda = \bigcap_{m\geq 1} \Lambda(K_m) \neq \phi$$

since all $\Lambda(K_m)$, $m \geq 1$ are compact subsets of a compact set K_1. Since $F^t\Lambda = \Lambda(K_m)$ for all $m \geq 1$ then $F^t\Lambda = \Lambda$ and by (3.5), $\Lambda \subset K$. Thus $\phi \neq \Lambda \subset \Lambda(K)$. This contradicts our assumption and we conclude that for some $m_0 \geq 1$, $\Lambda(K_{m_0}) = \phi$. It remains to put $\tilde{K} = K_{m_0}$.

Next, for any $x \in \tilde{K}$ put

$$t(x) = \inf\{t \geq 0 : F^t x \notin \tilde{K}\}. \tag{3.6}$$

We claim that

$$t(x) < \infty \quad \text{for each} \quad x \in \tilde{K}. \tag{3.7}$$

Indeed, if $t(x_0) = \infty$ then $F^t x_0 \in \tilde{K}$ for all $t \geq 0$. Thus for some sequence $t_n \uparrow \infty$ and a point $y \in \tilde{K}$ one has $F^{t_n} x_0 \to y$ as $t_n \to \infty$. Then also $F^{t_n+t} x_0 \to F^t y$ as $t_n \to \infty$ for any $t \in (-\infty,\infty)$. Thus $F^t y \in \tilde{K}$ for all

$t \in (-\infty,\infty)$, and so the set $\{F^t y, t \in (-\infty,\infty)\} \subset \Lambda(\tilde{K}) = \phi$. This contradiction proves (3.7). Notice that the function $t(x)$ is upper semicontinuous since if $F^t x \notin \tilde{K}$, i.e., $F^t x$ belongs to the open set $M\setminus\tilde{K}$ then $F^t y \in M\setminus\tilde{K}$ for any y close enough to x, and so $\limsup_{y\to x} t(y) \leq t(x)$. This implies

$$L = \sup_{x\in\tilde{K}} t(x) < \infty. \tag{3.8}$$

By Theorem 1.2 (i) we conclude that

$$\lim_{\epsilon\to 0} \sup_{x\in K} P_x^\epsilon\{\text{dist}(X_t^\epsilon, F^t x) > \delta\} = 0 \tag{3.9}$$

for any $\delta > 0$ and $t > 0$. Now (3.6)-(3.9) imply

$$\rho(\epsilon) = \sup_{x\in K} P_x^\epsilon\{\tau > L\} \to 0 \quad \text{as} \quad \epsilon \to 0. \tag{3.10}$$

By the Markov property (see, for instance, Doob [Do] or Friedman [Fri], vol.1),

$$\begin{aligned} P_x^\epsilon\{\tau > mL\} &= E_x^\epsilon \chi_{\tau>L} E_{X_L^\epsilon}^\epsilon \chi_{\tau>L} \cdots E_{X_{(m-1)L}^\epsilon}^\epsilon \chi_{\tau>L} \\ &\leq (\rho(\epsilon))^m, \end{aligned} \tag{3.11}$$

where χ_A is the indicator of the event A. Since $P_x^\epsilon\{\tau > t\}$ decreases in t we obtain

$$Q_x^\epsilon = \limsup_{t\to\infty} \frac{1}{t} \log P_x^\epsilon\{\tau > t\} \leq \frac{1}{L} \log \rho(\epsilon). \tag{3.12}$$

Taking into account (3.10) we derive from (3.12) that for any $x \in K$,

$$Q_x^\epsilon \to -\infty \quad \text{as} \quad \epsilon \to 0 \tag{3.13}$$

which contradicts (3.3). Hence the assumption $\Lambda(K) = \phi$ is inconsistent.

To prove the item (b) notice that

$$E_x^\epsilon \tau \leq L \sum_{m=1}^{\infty} P_x^\epsilon\{\tau > (m-1)L\}. \tag{3.14}$$

Assumption $\Lambda(K) = \phi$ gives, in view of (3.10), (3.11) and (3.14) for ϵ small enough and $x \in K$, that

$$E_x^\epsilon \tau \leq L \sum_{m=1}^{\infty} (\rho(\epsilon))^{m-1} < \infty. \tag{3.15}$$

This contradiction with (3.4) completes the proof of Theorem 3.1. □

1.4. Attractors and limiting measures.

In this section which follows partly the paper of Ruelle [Ru5] we shall introduce attractors and stable invariant sets and show that limiting measures of local random perturbations must have support on these sets.

We shall consider a metric space M together with a family F^t, $t \in T$ of continuous maps $F^t: M \to M$ where the set of indices T coincides with either $\mathbb{Z} = \{0,\pm1,\pm2,\dots\}$, $\mathbb{Z}_+ = \{0,1,2,\dots\}$ (discrete time case), or $\mathbb{R} = (-\infty,\infty)$, $\mathbb{R}_+ = [0,\infty)$ (continuous time case). The group or semigroup properties $F^0 =$ identity (= id), $F^s \circ F^t = F^{s+t}$ hold true whenever $s,t \in T$. A sequence of points $x_0,\dots,x_n \in M$ will be called a δ-pseudo-orbit of length n if

$$\operatorname{dist}(F^1 x_i, x_{i+1}) < \delta \quad \text{for all} \quad i = 0,\dots,n-1. \tag{4.1}$$

Any single point sequence will be considered as a δ-pseudo-orbit, as well.

For a pair of points $x, y \in M$ we shall write $x \to y$ if for any $\delta > 0$ there exist a number $t \in T$, $0 \leq t < 1$ and a δ-pseudo-orbit $x_0, \dots, x_n$ such that $F^t x = x_0$ and $x_n = y$. If $T = \mathbb{Z}$ or $T = \mathbb{Z}_+$ then, clearly, $t = 0$ here. Furthermore, we shall write $z \succ x$ if there exists a sequence of points $y_0, \dots, y_k \in M$ such that $y_0 = x$, $y_k = z$ and $y_i \to y_{i+1}$ for all $i = 0, 1, \dots, k-1$. Clearly, the relation "$\succ$" is reflexive ($x \succ x$) and transitive ($x \succ y$ and $y \succ z$ imply $x \succ z$). If $x \succ y$ and $y \succ x$ we shall write $x \sim y$. Evidently, "$\sim$" is an equivalence relation. As usual, any maximal set of equivalent points in M will be called an equivalence class. One concludes from the definition that each equivalence class is a closed set. An equivalence class containing a point x will be denoted by $[x]$. An equivalence class $[x]$ is called a basic class if either $F^t x = x$ for all $t \in T$ or $[x]$ contains more - than one point. Non-basic equivalence classes are not interesting but we have to admit them to comply with the tradition saying that an equivalence relation must be reflexive. Remark that our equivalence relation is slightly different from Ruelle's [Ru5].

The relation "$\succ$" induces a partial order on the equivalence classes so that $[x] \succ [y]$ if $x \succ y$ which is, clearly, well defined. A maximal (in this partial order) equivalence class will be called a quasiattractor. Hurley [Hu] calls this maximal equivalence classes more extensively: chain transitive quasiattractors. Ruelle [Ru5] suggests to call them simply attractors. We shall adopt here more common definition saying that a closed set $\Lambda \subset M$ is an attractor if it has an open neighborhood U such that

$$\bigcap_{t \in T,\, t \geq 0} F^t U = \Lambda \quad \text{and} \quad F^t U \subset V,\ t \in T \tag{4.2}$$

for every open set $V \supset \Lambda$ provided $t \geq t(V)$ is large enough. A set U satisfying (4.2) is called a fundamental neighborhood of the attractor Λ. The open set $W = \bigcup_{t \in T} (F^t)^{-1} U$ is called the basin of attraction of Λ. The set W consists of $x \in M$ such that $F^t x \to \Lambda$ as $t \to \infty$, and so W is independent of the choice of U. If $\{F^t, t \in T\}$ is a group we may take Λ equal to the whole space M, and then $\Lambda = U = W$. Notice that the union of basic equivalence classes is Conley's chain recurrent set, which can be described by means of attractors (see Conley [Con], p.37). For other relations between basic equivalence classes, attractors and quasiattractors we refer the reader to Ruelle [Ru5].

One can give the following characterization of basic equivalence classes.

Proposition 4.1. *An equivalence class* $[x]$ *is basic if and only if*

$$F^t[x] \subset [x] \quad \textit{for all} \quad t \in T. \tag{4.3}$$

Proof. If $[x]$ is not basic then x is the only point belonging to $[x]$ and x is not a fixed point. Hence for some $t \in T$, $x \neq F^t x \notin [x]$ and so (4.3) fails. It remains to show that if $[x]$ is basic then (4.3) holds true. If $[x]$ is a single point set and x is a fixed point, i.e., $F^t x = x$ for all $t \in T$, then, clearly, (4.3) is satisfied. Suppose now that $[x]$ contains another point $y \neq x$. Then for any $\delta > 0$ there exist integers $n_i \geq 0$, real numbers $t_i \in T$, $0 \leq t_i < 1$ and δ-pseudo-orbits $z_0^{(i)}, \ldots, z_{n_i}^{(i)}$, $i = 1, \ldots, k(\delta), \ldots, \ell(\delta)$ such that $F^{t_1} x = z_0^{(1)}$, $F^{t_{i+1}} z_{n_i}^{(i)} = z_0^{(i+1)}$, $i = 1, \ldots, \ell(\delta)-1$, $z_{n_{k(\delta)}}^{(k(\delta))} = y$ and $z_{n_{\ell(\delta)}}^{(\ell(\delta))} = x$. It suffices to prove (4.3) when $|t| \leq 1$. Suppose first that $t \geq 0$. Then $\tilde{x} = F^t x > x$. If $t \leq t_1$ then the same sequence of

δ-pseudo-orbits leads from $\tilde{x}$ to y. For $t > t_1$ consider $\tilde{z}_1 = F^{1+t_1-t}\tilde{x} = F^1 x_0^{(1)} = F^{1+t_1}x$. Since $\text{dist}(\tilde{z}_1, z_1^{(1)}) = \text{dist}(F^1 z_0^{(1)}, z_1^{(1)}) < \delta$ then in view of the continuity of $F^s y$ in (s,y) we conclude that there exists $\gamma_x(\delta) \to 0$ as $\delta \to 0$ such that

$$\sup_{0\leq u\leq 1, z\in U_\delta(F^{1+u}x)} \text{dist}(F^1 z, F^{1+u}x) < \gamma_x(\delta). \tag{4.4}$$

Thus $\text{dist}(F^1\tilde{z}_1, z_2^{(1)}) \leq \text{dist}(F^1 z_1^{(1)}, z_2^{(1)})$ $+ \text{dist}(F^1 z_1^{(1)} F^1\tilde{z}_1) < \delta + \gamma_x(\delta)$ and so $\tilde{z}_1, z_2^{(1)}, \ldots, z_{n_1}^{(1)}$ is a $(\delta+\gamma_x(\delta))$-pseudo-orbit. Since this can be done for any $\delta > 0$ and $\delta + \gamma_x(\delta) \to 0$ as $\delta \to 0$ we obtain $y \succ \tilde{x}$. This together with $\tilde{x} \succ x$ and $x \succ y$ give $\tilde{x} \sim x$, i.e., $\tilde{x} = F^t x \in [x]$. For negative $t \in T$ the proof is similar but we change the last δ-pseudo-orbit $z_0^{(\ell(\delta))}, \ldots, z_{n_{\ell(\delta)}}^{(\ell(\delta))}$ replacing $z_{n_{\ell(\delta)}-1}^{(\ell(\delta))}$ by $\tilde{z}_{n_{\ell(\delta)}-1}$ $= F^{-1}x$. Then $z_0^{(\ell(\delta))}, \ldots, z_{n_{\ell(\delta)}-2}^{(\ell(\delta))}, \tilde{z}_{n_{\ell(\delta)}-1}$ will be a $(\delta+\tilde{\gamma}_x(\delta))$-pseudo-orbit with $\tilde{\gamma}_x(\delta) \to 0$ as $\delta \to 0$. Finally, we add the length one δ-pseudo-orbit consisting of the point $F^t x$. Since $t < 0$, $F^{1+t}\tilde{z}_{n_{\ell(\delta)}-1} = F^t x$, and this construction goes through for any $\delta > 0$ we obtain $F^t x \succ y$. On the other hand, clearly, $x \succ F^t x$ (recall $t < 0$!) and $y \succ x$. Therefore $x \sim F^t x$, i.e., $F^t x \in [x]$ completing the proof. □

Corollary 4.1. *The union of all basic equivalence classes is a closed set. Any quasiattractor is a basic class.*

Proof is easy and we leave it to the reader.

The following result claims that under certain compactness and disjointness assumptions (which are satisfied, for instance, in the case of Smale's Axiom A dynamical systems: see Smale [Sm]) quasiattractors turn out to be attractors.

Proposition 4.2. *Let* $[x]$ *be a quasiattractor having an open neighborhood* $G \supset [x]$ *whose closure* $\bar{G}$ *is compact and* $\bar{G}$ *is disjoint from other basic equivalence classes except for* $[x]$. *Then* $[x]$ *is an attractor and*

$$F^t[x] = [x] \quad \textit{for any} \quad t \in T. \tag{4.5}$$

Proof. Since the family $\{F^t, 0 \leq t \leq 1\}$ is equicontinuous on $\bar{G}$, $[x] \subset G$ is a closed set, and by Proposition 4.1 $F^t[x] \subset [x]$ then there exists another open neighborhood W of $[x]$ such that $\bar{W} \subset G$ and $F^t\bar{W} \subset G$ for all $t \in [0,1]$. We shall call a δ-chain any family of δ-pseudo-orbits $z_0^{(i)}, \dots, z_{n_i}^{(i)}$, $i = 1, \dots, k(\delta)$ such that $F^{t_{i+1}} z_{n_i}^{(i)} = z_0^{(i+1)}$, $i = 1, \dots, k(\delta)-1$ for some integers $n_i \geq 0$ and real numbers $t_i \in T$, $0 \leq t_i \leq 1$. Clearly, $z \succ y$ if and only if for any $\delta > 0$ there is a δ-chain ζ leading from y to z, i.e., in the above definition $z_0^{(1)} = y$ and $z_{n_{k(\delta)}}^{(k(\delta))} = z$. We claim that there exists $\delta_0 > 0$ such that any δ_0-chain starting in a point $y \in [x]$ has all its points in W. Indeed, if this were not true then there would exist a sequence of numbers $\delta_n \downarrow 0$ and a sequence of δ_n-chains ζ_n which start at points $y_n \in [x]$ and end at points $z_n \in \bar{G} \backslash W$. Both sets $[x]$ and $\bar{G} \backslash W$ are compact, and so we can choose converging subsequences $y_{n_i} \to y \in [x]$ and $z_{n_i} \to z \in \bar{G} \backslash W$ as $i \to \infty$. Then we shall have that for any $\delta > 0$ there is a δ-chain leading from y to z, and so $z \succ y$. This contradicts the fact that $[x]$ is a quasiattractor. Now we are ready to establish (4.5). We know already by Proposition 4.1 that $F^t[x] \subset [x]$ for all $t \in T$. This already implies (4.5) if F^t, $t \in T$ is a group. It suffices to get (4.5) for $t = 1$ since by (4.3), $F^t[x] \supset F^1[x] = [x]$ provided $0 \leq t \leq 1$. Thus for any $y \in [x]$ we have to find $z \in [x]$ such that $F^1 z = y$. If $[x]$ is a fixed point then (4.5) is clear. If $[x]$ is not a single point then we can choose $\tilde{y} \in [x]$, $\tilde{y} \neq y$ such that for any sequence $\delta_n \downarrow 0$ there exists a sequence of δ_n-chains ζ_n which begin at $\tilde{y}$, end at y and contain points $v_n, w_n \in W$ satisfying

$$\operatorname{dist}(F^1 v_n, w_n) < \delta_n \quad \text{and} \quad F^{t_n} w_n = y$$

for some nonnegative $t_n < 1$. Choosing a subsequence n_i so that $t_{n_i} \to t_0$, $v_{n_i} \to v$ and $w_{n_i} \to w$ we obtain that $\tilde{y} \succ v \succ w$, and so $v, w \in [x]$ since $[x]$ is a quasiattractor. Clearly, $F^{1+t_0} v = y$, and $z = F^{t_0} v \in [x]$ by Proposition 4.1. Since $F^1 z = y$ then (4.5) is proved.

We already know that any δ_0-chain starting in $[x]$ remains entirely in W. This together with (4.5) imply that

$$\text{if } z \in U_0 = \{y : \operatorname{dist}(y, [x]) < \delta_0\} \text{ then} \tag{4.6}$$

$$F^t z \in W \text{ for all } t \geq 0.$$

In the same way as above one can see that there exists $\delta_1 > 0$, $\delta_1 < \delta_0$ such that

$$\text{if } z \in U_1 = \{y : \operatorname{dist}(y, [x]) < \delta_1\} \text{ then} \tag{4.7}$$

$$F^t z \in U_0 \text{ for all } t \geq 0.$$

Take an arbitrary open set $V \supset [x]$. It remains to show that $F^t U_1 \subset V$ provided $t \geq t(V)$ is large enough. Pick up an open set $V_1 \supset [x]$ such that $\bar{V}_1 \subset V$. We claim that

$$t(V) = \inf\{t : F^t U_0 \subset V_1\} < \infty. \tag{4.8}$$

Indeed, if (4.8) fails then there exist a sequence of points $z_n \in U_0$ and numbers $t_n \to \infty$ such that $F^t z_n \notin V_1$ for all $t \in [0, t_n]$. Choose a converging subsequence $z_{n_i} \to z \in \overline{U}_0$. Then $F^t z \notin V_1$ for all $t \geq 0$. On the other hand, since $F^t U_0 \subset W$ then $F^t \overline{U}_0 \subset \overline{W}$, and so $F^t z \in \overline{W}$ for all $t \geq 0$. Thus the whole trajectory $\{F^t z, t \geq 0\}$ stays in the compact set $\overline{W} \setminus V_1$ which is impossible since $\overline{W} \setminus V_1$ is disjoint from basic equivalence classes. Finally, from (4.7) and (4.8) it follows that for $t \geq t(V)$,

$$F^t U_1 = F^{t(V)} F^{t-t(V)} U_1 \subset F^{t(V)} U_0 \subset V \tag{4.9}$$

saying that $[x]$ is an attractor. □

Usually, the proofs are easier for discrete time dynamical systems. We shall often pass in this book from a semiflow (flow) F^t to its time-one map $F = F^1$. Of course, if we are dealing with the discrete time case ($T = \mathbb{Z}$ or $T = \mathbb{Z}_+$) nothing changes. Considering F^t, $t \in T$ only in integer times we pass to the discrete time dynamical system F^n, $n \in \mathbb{Z}$ or $n \in \mathbb{Z}_+$. For this discrete time dynamical system in the same way as above we can define the equivalence relation which we shall denote by "$\overset{1}{\sim}$". The corresponding equivalence classes we denote by $[\;]^{(1)}$.

Proposition 4.3. *For any* $x \in M$, $[x]^{(1)} \subset [x]$. *If* $[x]^{(1)}$ *is a basic class for* F^n *then* $[x]$ *is a basic class for* F^t.

Proof. The first assertion is obvious. Now let $[x]^{(1)}$ be a basic class. If $[x]^{(1)}$ contains another point y then so does $[x]$. If $[x]^{(1)}$ is the singleton then $F^1 x = x$ and so x is a periodic point for F^t. Then the whole periodic orbit $\{F^t x, 0 \leq t \leq 1\}$ belongs to $[x]$. □

Put $F = F^1$. A point $x \in M$ is called wandering for F if there is an open neighborhood U of x such that $U \cap F^{-n}U = \phi$ for all $n = 1,2,\ldots$, or, which is the same, the sets $F^{-n}U$, $n \geq 0$ are mutually disjoint. The nonwandering set $\Omega(F)$ for F consists of all points that are not wandering for F, i.e., $\Omega(F) = \{x \in M$: for every neighborhood U of x there is $n \geq 1$ with $F^{-n}U \cap U \neq \phi\}$.

Proposition 4.4. (a) $\Omega(F)$ *is closed;*

(b) $F\Omega(F) \subset \Omega(F)$;

(c) *If* F *is a homeomorphism then* $\Omega(F^{-1}) = \Omega(F)$ *and* $F\Omega(F) = \Omega(F)$;

(d) *If* $x \in \Omega(F)$ *then* $[x]^{(1)}$ *is a basic equivalence class;*

(e) *If* $\mu \in \mathcal{P}(M)$ *is F-invariant then* $\mu(\Omega(F)) = 1$ *provided* M *is a separable metric space.*

Proof. (a) From the definition of $\Omega(F)$ it is clear that $M \backslash \Omega(F)$ is open.

(b) Let $x \in \Omega(F)$ and let V be a neighborhood of Fx. Then $F^{-1}V$ is a neighborhood of x and so there is some $n > 0$ with $F^{-(n+1)}V \cap F^{-1}V \neq 0$. Thus $F^{-n}V \cap V \neq \phi$ implying $Fx \in \Omega(F)$.

(c) If F is homeomorphism then $F^{-1}U \cap U = F^{-n}(U \cap F^n U)$ and so $\Omega(F) = \Omega(F^{-1})$. But then applying (b) to F^{-1} we have $F^{-1}\Omega(F) \subset \Omega(F)$.

(d) If $x \in \Omega(F)$ then it follows easily from the definition that $x \overset{1}{\sim} Fx$, i.e., $Fx \in [x]^{(1)}$. If F is invertible then also $F^{-1}x \in [x]^{(1)}$. By Proposition 4.1 this implies that $[x]^{(1)}$ is a basic class.

(e) Let $\{U_n\}_1^\infty$ be a base for the topology. Then $M \setminus \Omega(F)$ is the union of those U_n such that the sets U_n, $F^{-1}U_n$, $F^{-2}U_n, \dots$ are pairwise disjoint. Such a set U_n must have measure zero for any F-invariant measure and so $\mu(M \setminus \Omega(F)) = 0$. □

Since every F^t-invariant measure $\mu \in \mathcal{P}(M)$ is also F^1-invariant then taking into account Theorem 1.1, Corollary 4.1, Propositions 4.3 and 4.4 we immediately derive

Corollary 4.2. *Let* F^t, $t \in T$ *be a family of continuous maps as above on a separable metric space* M. *Any* F^t*-invariant measure* $\mu \in \mathcal{P}(M)$ *has support in* $\Omega(F^1)$ *which is contained in the union of basic classes. In particular, if* ν^ϵ *are invariant measures of random perturbations* X_t^ϵ *of* F^t *such that* $\mu^{\epsilon_i} \to \nu$ *as* $\epsilon_i \to 0$. *Then* $\operatorname{supp} \nu \subset \Omega(F^1)$.

Next we shall study in details the case of localized random perturbations of a continuous map $F: M \to M$ of a metric space M. We shall consider a Markov chain X_n^ϵ with transition probabilities $P^\epsilon(x, \Gamma) = Q_{Fx}^\epsilon(\Gamma)$ where $Q_y^\epsilon \in \mathcal{P}(M)$, $y \in M$ satisfy

$$FU_\delta(y) \subset \operatorname{supp} Q_{Fy}^\epsilon \subset U_\epsilon(Fy) \tag{4.10}$$

for all $y \in M$ where $\delta = \delta(\epsilon) < \epsilon$ is independent of y, and, again, $U_\gamma(z) = \{v : \operatorname{dist}(v, z) < \gamma\}$. We assume also that Q_x^ϵ depends continuously on x in the topology of weak convergence on $\mathcal{P}(M)$, i.e., for any bounded $g \in C(M)$ the function $\int g(y) Q_x^\epsilon(dy)$ is continuous in x.

Theorem 4.5. (cf. Ruelle [Ru5]). *Let* $F: M \to M$ *be as above and suppose that an equivalence class* $[x]$ *is not a* *quasiattractor. Assume that transition probabilities of random perturbations* X_n^ϵ *satisfy (4.10) and* Q_x^ϵ *depends continuously on* x. *If* $\epsilon > 0$ *is small enough then for any measure* $\mu \in \mathcal{P}(M)$ *one has*

$$\lim_{n\to\infty}(P_\epsilon^*)^n\mu(U_\gamma(x)) = 0 \tag{4.11}$$

provided $\gamma > 0$, $\gamma \leq \gamma(\epsilon)$ *is small enough. Hence, if* μ_∞^ϵ *is a weak limit of measures* $(P_\epsilon^*)^n\mu$, *then* $x \notin \operatorname{supp}\mu_\infty^\epsilon$. *In particular, this will be the case if* $\mu = \mu^\epsilon$ *is* P_ϵ^**-invariant. In this case if* $\mu^{\epsilon_i} \xrightarrow{w} \tilde{\mu}$ *for some subsequence* $\epsilon_i \to 0$ *then* $\tilde{\mu}(U_\gamma(x)) = 0$.

Proof. Since $[x]$ is not a quasiattractor there is $y \succ x$ and $y \notin [x]$. This means that if $\epsilon > 0$ is small enough there exist no 2ϵ-pseudo-orbits going from y to x though there is a δ-pseudo-orbit $(z_0, z_1, \dots, z_m)$ with $z_0 = x$ and $z_m = y$. Let $z \in U_\delta(x)$. Then $Fz_0 = Fz \in FU_\delta(z)$ and so by (4.10), $Fz_0 \in \operatorname{supp} Q_{Fz}^\epsilon$. By induction one obtains

$$Fz_i \in \operatorname{supp}(P_\epsilon^*)^{i+1}\delta_z \quad \text{for all} \quad i = 0,\dots,m \tag{4.12}$$

where δ_z is the unit measure concentrated at z and $(P_\epsilon^*)^\ell\delta_z(\Gamma) = P^\epsilon(\ell, z, \Gamma)$ is the ℓ-step transition probability of X_n^ϵ. To prove (4.12) we remark that for any measure $\nu \in \mathcal{P}(M)$ one has

$$\operatorname{supp} P_\epsilon^*\nu = \text{closure} \bigcup_{v\in \operatorname{supp}\nu} \operatorname{supp} P_\epsilon^*\delta_v. \tag{4.13}$$

Indeed, since $P_\epsilon^*\nu(G) = \int d\nu(v)P^\epsilon(v,G) = \int d\nu(v)P_\epsilon^*\delta_v(G)$ then for any closed set G, $P_\epsilon^*\nu(G) = 1$ if and only if $P_\epsilon^*\delta_v(G) = 1$ for ν-almost all v. If $v_k \to v$ then

$\limsup_{k\to\infty} P^*_\epsilon \delta_{v_k}(G) \le P^*_\epsilon \delta_v(G)$ since G is closed and $P^*_\epsilon \delta_{v_k} \xrightarrow{w} P^*_\epsilon \delta_v$ by the assumption (see Billingsley [Bi], Theorem 2.1). Thus $P^*_\epsilon \nu(G) = 1$ if and only if $P^*_\epsilon \delta_v(G) = 1$ for all $v \in \operatorname{supp} \nu$ proving (4.13). Now assuming (4.12) to be true for $i = k-1$ we derive from (4.10) and (4.13) that

$$\operatorname{supp}(P^*_\epsilon)^{k+1}\delta_z = \text{closure} \bigcup_{v \in \operatorname{supp}(P^*_\epsilon)^k \delta_z} \operatorname{supp} P^*_\epsilon \delta_v \tag{4.14}$$

$$\supset \operatorname{supp} P^*_\epsilon \delta_{Fz_{k-1}} \supset FU_\delta(Fz_{k-1}) \supset Fz_k$$

proving (4.12) by induction.

Next, applying (4.12) for $i = m$ we conclude that

$$(P^*_\epsilon)^{m+1}\delta_z(U_\epsilon(Fy)) > 0 \quad \text{for any} \quad z \in U_\delta(x). \tag{4.15}$$

Since $(P^*_\epsilon)^{m+1}\delta_z$ is continuous in z with respect to the topology of weak convergence on $\mathcal{P}(M)$ we derive that there is $\gamma > 0$, $\gamma \le \delta$ such that

$$(P^*_\epsilon)^{m+1}\delta_z(U_\epsilon(Fy)) \ge \beta > 0 \quad \text{for all} \quad z \in U_\gamma(x) \tag{4.16}$$

where $\beta = \frac{1}{2}(P^*_\epsilon)^{m+1}\delta_x(U_\epsilon(Fy))$. Here we have used another basic fact about weak convergence saying that $\liminf_{n\to\infty} \nu_n(V) \ge \nu(V)$ provided $\nu_n \xrightarrow{w} \nu$ and V is an open set (see Billingsley [Bi], Theorem 2.1).

By the Chapman-Kolmogorov formula (1.19) and the right hand side of (4.10) we have for any $v \in M$,

$$1 = (P_\epsilon^*)^n \delta_v(M) \tag{4.17}$$

$$= \int_M \cdots \int_M Q_{Fv}^\epsilon(dv_1) Q_{Fv_1}^\epsilon(dv_2) \cdots Q_{Fv_{n-1}}^\epsilon(dv_n)$$

$$= \int_{U_\epsilon(Fv)} \int_{U_\epsilon(Fv_1)} \cdots \int_{U_\epsilon(Fv_{n-1})} Q_{Fv}^\epsilon(dv_1) Q_{Fv_1}^\epsilon(dv_2)$$

$$\times \cdots \times Q_{Fv_{n-1}}^\epsilon(dv_n).$$

Thus one obtains non-zero contributions in (4.17) only by integrating over ϵ-pseudo-orbits. Since by the assumption there is no 2ϵ-pseudo-orbit going from y to x then for all $n = 0,1,2,\ldots$

$$\mathrm{supp}(P_\epsilon^*)^n \delta_v \cap U_\delta(x) = \phi \quad \text{provided} \quad v \in U_\epsilon(Fy), \tag{4.18}$$

and so $P(n,v,U_\delta(x)) = 0$. Moreover for all $v \in U_\epsilon(Fy)$,

$$P_v\{X_n^\epsilon \in U_\delta(x) \text{ for some } n \geq 0\} \tag{4.19}$$

$$\leq \sum_{n=0}^{\infty} P(n,v,U_\delta(x)) = 0$$

where $P_v\{A\}$ denotes the probability of an event A for the Markov chain X_n^ϵ under the condition $X_0^\epsilon = v$. From (4.16), (4.17) and the Markov property (see, for instance Doob [Do]) it follows that for any $z \in U_\gamma(x)$,

$$P_z\{X_n^\epsilon \in U_\delta(x) \text{ for some } n \geq m+1\} \tag{4.20}$$

$$= \int_{M \setminus U_\epsilon(Fy)} P(m+1,z,dv) P_v\{X_n^\epsilon \in U_\delta(x) \text{ for some } n \geq 0\}$$

$$\leq P(m+1,z,M \setminus U_\epsilon(Fy)) \leq 1-\beta.$$

Next, we intend to show that for any $v \in M$,

$$P_v\{X_n^\epsilon \text{ visits } U_\gamma(x) \text{ infinitely many times}\} = 0. \quad (4.21)$$

Indeed, define inductively $\tau_1 = \inf\{n \geq m+1 : X_n^\epsilon \in U_\gamma(x)\}$ and $\tau_{i+1} = \inf\{n \geq \tau_i + m+1 : X_n^\epsilon \in U_\gamma(x)\}$ for $i = 1,2,\ldots$. All τ_i, $i = 1,2,\ldots$ are the Markov times (stopping times), and so by the strong Markov property (see Doob [Do] or Friedman [Fri], vol.1) for any $v \in M$ we derive from (4.20) that

$$P_v\{X_n^\epsilon \text{ visits } U_\gamma(x) \text{ more than } k(m+1) \text{ times}\} \quad (4.22)$$

$$\leq P_v\{\tau_k < \infty\} = E\chi_{\tau_1<\infty} \cdots \chi_{\tau_k<\infty}$$

$$= E_v \chi_{\tau_1<\infty} E_{X_{\tau_1}^\epsilon} \chi_{\tau_1<\infty} \cdots E_{X_{\tau_{k-1}}^\epsilon} \chi_{\tau_1<\infty}$$

$$\leq \Big(\sup_{z\in U_\gamma(x)} E_z \chi_{\tau_1<\infty}\Big)^{k-1}$$

$$= \Big(\sup_{z\in U_\gamma(x)} P\{X_n^\epsilon \in U_\gamma(x) \text{ for some } n \geq m+1\}\Big)^{k-1}$$

$$\leq (1-\beta)^{k-1}.$$

Since $\Sigma(1-\beta)^{k-1}$ converges then (4.22) together with the Borel-Cantelli lemma gives (4.21). Thus $P_v\{X_n^\epsilon$ visits $U_\gamma(x)$ for some $n \geq N\} \to 0$ as $N \to \infty$, and so $P(N,v,U_\gamma(x)) \to 0$ as $N \to \infty$. Therefore, by the dominated convergence theorem

$$(P_\epsilon^*)\mu(U_\gamma(z)) = \int_M d\mu(v)P^\epsilon(n,v,U_\gamma(x)) \to 0 \tag{4.23}$$

$$\text{as } n \to \infty$$

for any probability measure μ proving (4.11). Now let $\mu = \mu^\epsilon$ be P_ϵ^*-invariant then (4.23) gives $\mu^\epsilon(U_\gamma(x)) = 0$. If $\mu^{\epsilon_i} \xrightarrow{w} \tilde{\mu}$ for some subsequence $\epsilon_i \to 0$ then $0 = \liminf_{i\to\infty} \mu^{\epsilon_i}(U_\gamma(x)) \geq \tilde{\mu}(U_\gamma(x))$ (see Billingsley [Bi], Theorem 2.1) and so $\tilde{\mu}(U_\gamma(x)) = 0$. □

1.5. Attractors and limiting measures via large deviations.

In this section we shall introduce a machinery which enables one to treat certain nonlocalized random perturbations in the spirit of the previous section. An application of these to the continuous time case generalizes corresponding results of Wentzell and Freidlin (see [WF] and [FW]).

We shall consider again a metric space M together with a continuous map $F:M \to M$. In this section random perturbations X_n^ϵ defined by (1.1)-(1.2) will satisfy the following large deviations condition.

Assumption 5.1. There exists a continuous function $\rho(x,y)$ of $x,y \in M$ such that $\rho(x,y) > 0$ if $y \neq Fx$ and uniformly in $x \in M$,

$$\lim_{\epsilon\to 0} \epsilon \log Q_{Fx}^\epsilon(U) = -\inf_{y\in U} \rho(x,y) \tag{5.1}$$

for any open set $U \subset M$.

Remark that (1.1) implies $Q_{Fx}^\epsilon(U) \to 1$ as $\epsilon \to 0$ if $Fx \in U$, and so by (5.1) the function ρ must satisfy

$\rho(x,Fx) = 0$. One obtains an example of the above situation when $M = \mathbb{R}^m$, transition probabilities $P^\epsilon(x,\cdot) = Q^\epsilon_{Fx}(\cdot)$ have densities $p^\epsilon(x,y)$, i.e., $Q^\epsilon_{Fx}(U) = \int_U p^\epsilon(x,y)dy$, such that $\lim_{\epsilon\to 0} \epsilon \log p^\epsilon(x,y) = -\rho(x,y)$. If $\rho(x,y)$ is smooth except for $y = Fx$ one can use Laplace's method (see DeBruijn [D8]) to obtain (5.1). In particular, we may have $p^\epsilon(x,y) = (\pi\epsilon)^{-\frac{m}{2}} \exp\left[-\frac{|y-Fx|^2}{\epsilon}\right]$. In a more general case, the function $\rho(x,y)$ may be of the form $(\mathrm{dist}(Fx,y))^\alpha$ for some $\alpha > 0$.

Let A_N be a function on the N-fold product $M^N = M \times \ldots \times M$ defined for $\xi = (\xi_0,\ldots,\xi_{N-1}) \in M^N$, $\xi_i \in M$, $i = 0,\ldots,N-1$ by the formula

$$A_N(\xi) = \sum_{i=0}^{N-2} \rho(\xi_i,\xi_{i+1}) \text{ for } N > 1 \text{ and } A_1 \equiv 0. \qquad (5.2)$$

The function $A_N(\xi)$ (called, sometimes an action functional) measures the deviation of a sequence ξ from orbits of the map F. We shall see that when $\epsilon > 0$ is small the quantity $A_N(\xi)$ characterizes the difficulty for Markov chains X^ϵ_n to stay in a small tube near ξ during the first N steps. For any couple of points $x,y \in M$ we define also

$$B(x,y) \qquad (5.3)$$

$$= \inf\{A_n(\xi): n \geq 1, \xi = (\xi_0,\ldots,\xi_{n-1}), \xi_0 = x, \xi_{n-1} = y\}$$

The function B induces a partial order writing $y \succ x$ if $B(x,y) = 0$. This yields an equivalence relation saying $x \sim y$ if $x \succ y$ and $y \succ x$. Sometimes we shall denote these relations by $\overset{\rho}{\succ}$ and $\overset{\rho}{\sim}$ when dealing simultaneously with these relations and the relations defined in the previous section through δ-pseudo-orbits. Equivalence classes corresponding to $\overset{\rho}{\sim}$ we shall call ρ-equivalence clases. A ρ-equivalence class containing a point $x \in M$ will be denoted by $[x]_\rho$. A ρ-equivalence class $[x]_\rho$ is called a basic class if either $Fx = x$ or $[x]_\rho$ contains more than one point. As in the previous section, we obtain also a partial order among ρ-equivalence classes saying $[y]_\rho \succ [x]_\rho$ if $y \overset{\rho}{\succ} x$. A maximal (in this partial order) equivalence class will be called a ρ-quasiattractor.

In what follows we shall need continuity properties of the function $B(x,y)$ defined by (5.3).

Lemma 5.1. *The function* $B(x,y)$ *is upper semicontinuous in both variables. If* $\rho(x,y)$ *is uniformly continuous in* $x,y \in M$ *(in particular, if* M *is compact) then* B *is a continuous function, and so* ρ*-equivalence classes are closed sets.*

Proof. Let $x_n \to x$ and $y_n \to y$ as $n \to \infty$. For any $\delta > 0$ we can choose N_δ such that

$$B(x,y) \le A_{N_\delta}(\xi^\delta) \le B(x,y) + \delta \tag{5.4}$$

for some $\xi^\delta = (\xi_0^\delta, \ldots, \xi_{N_\delta - 1}^\delta)$ with $\xi_0^\delta = x$ and $\xi_{N_\delta - 1}^\delta = y$. Take now $\xi^{\delta,n}$ which coincides with ξ^δ everywhere except for the first and the last points which are chosen to be $\xi_0^{\delta,n} = x_n$ and $\xi_{N_\delta - 1}^{\delta,n} = y_n$. Since the function ρ is continuous we obtain by (5.2) and (5.4) that

$$B(x_n, y_n) \le A_{N_\delta}(\xi^{\delta,n}) \xrightarrow[n\to\infty]{} A_{N_\delta}(\xi^\delta) \le B(x,y) + \delta.$$

Since $\delta > 0$ can be chosen arbitrarily small this implies

$$\limsup_{n\to\infty} B(x_n,y_n) \le B(x,y), \tag{5.5}$$

i.e., B is upper semicontinuous.

Suppose now that $\rho(v,w)$ is uniformly continuous in both variables. For any $\delta > 0$ and an integer $n > 0$ we can choose $N_{\delta,n}$ such that

$$B(x_n,y_n) \le A_{N_{\delta,n}}(\eta^{\delta,n}) \le B(x_n,y_n) + \delta \tag{5.6}$$

for some $\eta^{\delta,n} = (\eta_0^{\delta,n},\dots,\eta_{N_{\delta,n}-1}^{\delta,n})$ with $\eta_0^{\delta,n} = x_n$ and $\eta_{N_{\delta,n}-1}^{\delta,n} = y_n$. Take now $\tilde{\eta}^{\delta,n}$ which coincides with $\eta^{\delta,n}$ everywhere except for the first and the last points where we put $\tilde{\eta}_0^{\delta,n} = x$ and $\tilde{\eta}_{N_{\delta,n}-1}^{\delta,n} = y$. Since ρ is uniformly continuous and $x_n \to x$, $y_n \to y$ then
$|\rho(x_n,\eta_1^{\delta,n})-\rho(x,\eta_1^{\delta,n})| + |\rho(\eta_{N_{\delta,n}-2}^{\delta,n},y_n)-\rho(\eta_{N_{\delta,n}-2}^{\delta,n},y)| \to 0$

as $n \to 0$, and so

$$|A_{N_{\delta,n}}(\eta^{\delta,n})-A_{N_{\delta,n}}(\tilde{\eta}^{\delta,n})| \xrightarrow[n\to 0]{} 0. \tag{5.7}$$

Notice that $B(x,y) \le A_{N_{\delta,n}}(\tilde{\eta}^{\delta,n})$. Since $\delta > 0$ is arbitrarily small this together with (5.6) and (5.7) imply

$$B(x,y) \le \liminf_{n\to\infty} B(x_n,y_n).$$

In view of (5.5) we conclude from here that $B(x,y)$ is a continuous function. In this case any set $\{y : B(x,y) = 0\}$ is closed, and so ρ-equivalence classes are closed sets. □

Remark 5.1. The reader can check that the uniform continuity condition on ρ in the above lemma and in what follows can be replaced by the following weaker condition:

for any $\delta, q > 0$ there exists $\epsilon > 0$ such that $|\rho(x,y)-\rho(\tilde{x},\tilde{y})| \le \delta$ whenever $\rho(x,y) \le q$ and $\mathrm{dist}(x,\tilde{x}) + \mathrm{dist}(y,\tilde{y}) \le \epsilon$. This last condition holds true, for instance, when $\rho(x,y) = (\mathrm{dist}(Fx,y))^n$, $n > 1$ and $M = \mathbb{R}^m$, while ρ fails to be uniformly continuous in this case.

In the remaining part of this section we suppose that both Assumption 5.1 and the following assumption are satisfied.

Assumption 5.2. The function ρ is uniformly continuous in both variables and the sets $\{y:\rho(x,y) \le a\}$ are compact for any $x \in M$ and $a \ge 0$.

Similarly to Proposition 4.2 we can compare the notions of ρ-quasiattractors and attractors.

Proposition 5.1. *Let* $[x]_\rho$ *be a* ρ*-quasiattractor having an open neighborhood* $G \supset [x]_\rho$ *whose closure* $\bar{G}$ *is compact and* $\bar{G}$ *is disjoint from other basic* ρ*-equivalence classes except for* $[x]_\rho$. *Then* $[x]_\rho$ *is an attractor and*

$$F[x]_\rho = [x]_\rho. \tag{5.8}$$

Proof. By Lemma 5.1 $[x]_\rho$ is closed, whence, compact set. Since $\rho(y,Fy) = 0$ for any $y \in M$ then $Fy \overset{\rho}{>} y$. For a ρ-quasiattractor $[x]_\rho$ this means that $F[x]_\rho \subset [x]_\rho$. By the continuity of F it follows that there exists an open set $W \supset [x]_\rho$ such that $\bar{W} \cup F\bar{W} \subset G$. We claim that there exists $\delta_0 > 0$ such that any sequence $\xi = (\xi_0,\ldots,\xi_{k-1})$ with $\xi_0 \in [x]_\rho$ and $A_k(\xi) \le \delta_0$ has all its points in W. Indeed, if this were not true then there would exist a sequence of numbers $\delta_n \downarrow 0$ and a collection of sequences $\xi^{(n)} = (\xi_0^{(n)},\ldots,\xi_{k_n-1}^{(n)})$ with $A_{k_n}(\xi^{(n)}) \le \delta_n$ which start at points $y_n = \xi_0^{(n)} \in [x]_\rho$

and end at points $z_n = \xi^{(n)}_{k_n-1} \notin W$. First, we shall show that z_n can be chosen to belong $\overline{G}\setminus W$ since for otherwise there would be no points of $\xi^{(n)}$ in $\overline{G}\setminus W$ which means that there would exist points $v_n \in W$ and $w_n \in M\setminus\overline{G}$ such that $\rho(v_n, w_n) \leq \delta_n$. Take a convergent subsequence $v_{n_i} \to v \in \overline{W}$ as $i \to \infty$. Since $F\overline{W} \subset G$ then for some $\gamma > 0$, $U_\gamma(Fv) = \{u : \mathrm{dist}(u, Fv) < \gamma\} \subset G$. By Assumption 5.2 all sets $K_\delta(v) = \{u : \rho(v,u) \leq \delta\}$ are compact. Clearly, $K_\delta(v)$ decreases when $\delta \downarrow 0$ and $\bigcap_{\delta \geq 0} K_\delta(v) = Fv$. It follows then that for some $\delta > 0$, $K_\delta(v) \subset U_\gamma(Fv) \subset G$. By the uniform continuity of ρ we conclude that $\rho(v_n, w) > \frac{\delta}{2}$ for any $w \in M\setminus\overline{G}$ provided n is big enough. This contradicts $\rho(v_n, w_n) \leq \delta_n$ for $w_n \in M\setminus\overline{G}$ and $\delta_n \downarrow 0$. Thus we can choose $z_n = \xi^{(n)}_{k_n-1} \in \overline{G}\setminus\overline{W}$. Both sets $[x]_\rho$ and $\overline{G}\setminus W$ are compact, which enables one to choose converging subsequences $y_{n_i} \to y \in [x]_\rho$ and $z_{n_i} \to z \in \overline{G}\setminus W$ as $i \to \infty$. Since the function ρ is uniformly continuous in $\overline{G}$ and $A_{k_n}(\xi^{(n)}) \leq \delta_n$ with $\delta_n \downarrow 0$ then we conclude that $B(y,z) = 0$. This contradicts the fact that $[x]_\rho$ is a ρ-quasiattractor since $z \notin [x]_\rho$, proving that $\xi_i \in W$ for all $i = 0, \ldots, k-1$ whenever $A_k(\xi) \leq \delta_0$ and $\xi_0 \in [x]_\rho$ for some small $\delta_0 > 0$ independent of ξ. In particular, we obtain that $D_{\delta_0} \subset W$ where $D_\delta = \{y : B(x,y) < \delta\}$. Moreover the same arguments show that

$$\bigcap_{\delta \geq 0} D_\delta = [x]_\rho. \tag{5.9}$$

Since we already know that $F[x]_\rho \subset [x]_\rho$ then to establish (5.8) we have to find for any $y \in [x]_\rho$ a point $z \in [x]_\rho$ such that $Fz = y$. If $[x]_\rho$ is a fixed point then (5.8) is clear. If $[x]_\rho$ is not a single point then we can choose $v \in [x]_\rho$ such that there exists a family $\xi^{(n)} = (\xi_0^{(n)},\dots,\xi_{k_n-1}^{(n)})$ such that $A(\xi^{(n)}) \leq \delta_n$, $\xi_0^{(n)} = v$, $\xi_{k_n-1}^{(n)} = y$ with $\delta_n \downarrow 0$. Taking a convergent subsequence $\xi_{k_{n_i}-2}^{(n_i)} \to w$ as $i \to 0$ we conclude by (5.9) that $w \in [x]_\rho$. Since $\rho(\xi_{k_{n_i}-2}^{(n_i)},y) \leq \delta_{n_i}$ then $\rho(w,y) = 0$, and so $Fw = y$ proving (5.8).

Notice that

$$B(z,w) \leq B(z,y) + B(y,w) \tag{5.10}$$

for any triple $z,y,w \in M$. In particular,

$$B(x,Fy) \leq B(x,y) + B(y,Fy).$$

Since $B(y,Fy) = 0$ this means that $FD_\delta \subset D_\delta$ for any $\delta > 0$. Put $U = D_{\delta_0} \subset W$. Take an arbitrary open set $V \supset [x]_\rho$. It remains to show that $F^nU \subset V$ provided $n \geq n(V)$ is large enough. We claim that

$$n(V) = \inf\{n: F^nU \subset V\} < \infty. \tag{5.11}$$

Indeed, if (5.11) fails then there exist a sequence of points $z_i \in U$ and numbers $n_i \to \infty$ such that $F^n z_i \notin V$ for all $n = 0,1,\ldots,n_i$. Take a convergent subsequence $z_{i_j} \to z \in \bar{U} \subset \bar{W}$. Then we shall have that $F^n z \notin V$ for all $n = 0,1,\ldots$. On the other hand, since $FU \subset U$ then $F\bar{U} \subset \bar{U}$, and so $F^n z \in \bar{U}$ for all $n = 0,1,\ldots$. This means that the whole orbit $\{F^n z, n \geq 0\}$ stays in the compact set $\bar{U} \backslash V$ which is impossible since $\bar{U} \backslash V$ is disjoint from basic equivalence classes. Thus $n(V) < \infty$ and for any $n \geq n(V)$ we have

$$F^n U = F^{n(V)} F^{n-n(V)} U \subset F^{n(V)} U \subset V.$$

Since V is arbitrary, then also $\bigcap_{n \geq 0} F^n U = [x]_\rho$, and so $[x]_\rho$ is an attractor. □

Proposition 5.1 has the following direct application to random perturbations X_n^ϵ with transition probabilities $P^\epsilon(x,\cdot) = Q_{Fx}^\epsilon(\cdot)$ satisfying (5.1).

Corollary 5.1. *Let* $K = [x]_\rho$ *be a* ρ*-quasiattractor satisfying conditions of Proposition 5.1. Then for any open set* $V \supset K$ *there exist numbers* $r, \beta, \epsilon_0 > 0$ *such that for all* $N = 1,2,\ldots$ *one has*

$$P_x^\epsilon\{\tau_{M \backslash V} < N\} < N^2 e^{-\frac{\beta}{\epsilon}} \tag{5.12}$$

provided $x \in U_r(K) = \{y : \mathrm{dist}(y,K) < r\}$, $0 < \epsilon < \epsilon_0$, *where* $\tau_W = \inf\{n : X_n^\epsilon \in W\}$. *In particular,*

$$E^{\epsilon}_x \tau_{M\setminus V} > \frac{1}{4} e^{\frac{\beta}{2\epsilon}} . \tag{5.13}$$

Proof. By the Markov property

$$P^{\epsilon}_x\{\tau_{M\setminus V} < N\} = \sum_{n=1}^{N-1} P^{\epsilon}_x\{\tau_{M\setminus V} = n\} \tag{5.14}$$

$$= \sum_{n=1}^{N-1} P^{\epsilon}_x\{X^{\epsilon}_n \in M\setminus V \text{ and } X^{\epsilon}_i \in V \text{ for all } i = 1,\dots,n-1\}$$

$$= \sum_{n=1}^{N-1} \int_U \cdots \int_U P^{\epsilon}(x,dz_1)P^{\epsilon}(z_1 dz_2)\cdots P^{\epsilon}(z_{n-2},dz_{n-1})$$

$$\times P^{\epsilon}(z_{n-1},M\setminus V)$$

$$\le \sum_{n=1}^{N-1} \int_{U_\delta(Fx)} \int_{U_\delta(Fz_1)} \cdots \int_{U_\delta(Fz_{n-2})} \int_{U_\delta(Fz_{n-1})\cap(M\setminus V)}$$

$$P^{\epsilon}(x,dz_1)P^{\epsilon}(z_1,dz_2)\cdots P^{\epsilon}(z_{n-1},dz_n)$$

$$+ \frac{N(N-1)}{2} \sup_{z\in V} P^{\epsilon}(z,M\setminus U_\delta(Fz)).$$

Remark that the integrals in the last expression in (5.14) are taken over δ-pseudo-orbits starting at $x \in U_r(K)$ and ending in $M\setminus V$. Since K is an attractor then in the same way as in the proof of Proposition 4.2 one can see that there is no such δ-pseudo-orbits provided $\delta, r > 0$ are small enough. Thus the sum of integrals in the right hand side of the inequality in (5.14) is equal to zero.

Since the left hand side of (5.12) may only become bigger if V is taken smaller then without loss of generality we shall choose V from the beginning to satisfy $\text{dist}(V \cup FV, M\backslash G) = \delta_0 > 0$ where $\bar{G}$ is compact. Taking into account Assumptions 5.1 and 5.2 we derive for $\delta < \delta_0$ that

$$P^{\epsilon}(z, M\backslash U_{\delta}(Fz)) \leq P^{\epsilon}(z, \bar{G}\backslash U_{\delta}(Fz)) + P^{\epsilon}(z, M\backslash\bar{G})$$

$$\leq \sum_{i:\text{dist}(y_i, Fz) \geq \frac{2}{3}\delta} P^{\epsilon}(z, U_{\frac{\delta}{3}}(y_i) \cap \bar{G}) + P^{\epsilon}(z, M\backslash\bar{G})$$

$$< e^{-\frac{\beta}{\epsilon}}$$

where $U_{\frac{\delta}{3}}(y_i)$ is a finite cover of $\bar{G}$,

$$\beta = \frac{1}{2}\min(\inf\{\rho(z,v): z \in V, v \in M\backslash\bar{G}\},$$

$$\inf\{\rho(z,w): z \in V, w \in U_{\frac{\delta}{3}}(y_i) \cap \bar{G}, \text{dist}(y_i, Fz) \geq \tfrac{2}{3}\delta) > 0,$$

and $\epsilon > 0$ is small enough. This together with (5.14) and the disappearance of integrals in the right hand side of (5.14) yield (5.12). We obtain (5.13) noting that

$$E_x^{\epsilon}\tau_{M\backslash V} \geq NP_x^{\epsilon}\{\tau_{M\backslash V} < N\} \geq N(1 - N^2 e^{-\frac{\beta}{\epsilon}}) \quad \text{for any } N = 1, 2, \ldots,$$

and so the desired estimate will be achieved for N of order $\frac{1}{3} e^{\frac{\beta}{2\epsilon}}$. □

The following result shows the connection between the functional $A_n(\xi)$ and the probability for Markov chains X_k^ϵ to stay in a small tube near ξ during the first n steps. For any two sequences $\xi = (\xi_0, \ldots, \xi_{n-1})$ and $\zeta = (\zeta_0, \ldots, \zeta_{n-1})$, $\xi_i, \zeta_i \in M$ we put $\mathrm{dist}_n(\xi, \zeta) = \max_{0 \le i \le n-1} \mathrm{dist}(\xi_i, \zeta_i)$ which is the distance on M^n.

Theorem 5.2. (a) *For any* $\delta, \beta, N > 0$ *there is* $\epsilon_0 > 0$ *such that if* $\epsilon < \epsilon_0$ *then*

$$P_x^\epsilon\{\mathrm{dist}_n(X^\epsilon, \xi) < \delta\} \ge \exp\left[-\frac{(A_n(\xi)+\beta)}{\epsilon}\right] \tag{5.15}$$

for any sequence $\xi = (\xi_0, \ldots, \xi_{n-1})$ *with* $\xi_0 = x$ *and* $n \le N$, *where* X^ϵ *denotes the sequence* $(X_0^\epsilon, \ldots, X_{n-1}^\epsilon)$.

(b) *For any* $\beta, N > 0$ *there exist* $\delta_0, \epsilon_0 > 0$ *such that if* $\delta < \delta_0$ *and* $\epsilon < \epsilon_0$ *then*

$$P_x^\epsilon\{\mathrm{dist}_n(X^\epsilon, \xi) < \delta\} \le \exp\left[-\frac{(A_n(\xi)-\beta)}{\epsilon}\right] \tag{5.16}$$

for any sequence $\xi = (\xi_0, \xi_1, \ldots, \xi_{n-1})$ *with* $\xi_0 = x$ *and* $n \le N$.

Proof. (a) Put

$$\eta_\gamma = \sup_{\mathrm{dist}(y,y') \le \gamma, \mathrm{dist}(z,z') \le \gamma} |\rho(y,z) - \rho(y',z')|. \tag{5.17}$$

By Assumption 5.2, $\eta_\gamma \to 0$ as $\gamma \to 0$. Let $\gamma \le \delta$ then by Assumption 5.1 and the Chapman-Kolmogorov formula one has

$$P_x^\epsilon\{\mathrm{dist}_n(X^\epsilon,\xi) < \delta\} \geq P_x^\epsilon\{\mathrm{dist}_n(X^\epsilon,\xi) < \gamma\} \tag{5.18}$$

$$= \int_{U_\gamma(\xi_1)} \cdots \int_{U_\gamma(\xi_{n-1})} P^\epsilon(x,dz_1)P^\epsilon(z_1,dz_2)\cdots P^\epsilon(z_{n-1},dz_{n-1})$$

$$\geq P^\epsilon(x,U_\gamma(\xi_1)) \prod_{i=1}^{n-2} \inf_{z\in U_\gamma(\xi_i)} P^\epsilon(z,U_\gamma(\xi_{i+1}))$$

$$\geq \exp(-\epsilon^{-1}(nh_\epsilon + \sup_{z\in U_\gamma(\xi_1)} \rho(x,\xi_1)))$$

$$\prod_{i=1}^{n-2} \inf_{z\in U_\gamma(\xi_i)} \exp(- \sup_{v\in U_\gamma(\xi_{i+1})} \rho(z,v))$$

$$\geq \exp(-\epsilon^{-1}(A_n(\xi) + n(h_\epsilon+\eta_\gamma)))$$

where $h_\epsilon \geq 0$, $h_\epsilon \to 0$ as $\epsilon \to 0$. Since $n \leq N$, $\gamma > 0$ can be chosen arbitrarily small, and $\eta_\gamma \to 0$ as $\gamma \to 0$ then $n(h_\epsilon+\eta_\gamma)$ can be made less than β yielding (5.15).

(b) Again by Assumption 5.1 and the Chapman-Kolmogorov formula it follows

$$P_x^\epsilon\{\mathrm{dist}_n(X_i^\epsilon,\xi_i) < \delta\} \tag{5.19}$$

$$= \int_{U_\delta(\xi_1)} \cdots \int_{U_\delta(\xi_{n-1})} P^\epsilon(x,dz_1)P^\epsilon(z_1,dz_2)\cdots P^\epsilon(z_{n-2},dz_{n-1})$$

$$\leq P^\epsilon(x,U_\delta(\xi_1)) \prod_{i=1}^{n-2} \sup_{z\in U_\delta(\xi_i)} P^\epsilon(z,U_\delta(\xi_{i+1}))$$

$$\leq \exp(-\epsilon^{-1}(\inf_{z\in U_\rho(\xi_1)} \rho(x,\xi_1)-nh_\epsilon))$$

$$\prod_{i=1}^{n-2} \sup_{z\in U_\rho(\xi_i)} \exp(- \inf_{v\in U_\rho(\xi_{i+1})} \rho(z,v))$$

$$\leq \exp(-\epsilon^{-1}(A_n(\xi)-n(h_\epsilon+\eta_\delta)))$$

with $h_\epsilon \xrightarrow[\epsilon\to 0]{} 0$. Taking $\epsilon > 0$ and $\delta > 0$ small enough we can make $n(h_\epsilon+\eta_\delta)$ less than β proving (5.16). □

The following exposition may proceed also for a noncompact case similar to one considered in Theorem 1.7. Still, in order to simplify the arguments we shall assume in the remaining part of this section that M is compact which supersedes Assumption 5.2.

Assumption 5.3. M is compact.

Corollary 5.2. *Let*

$$\Phi_x(n,s) = \{\xi = (\xi_0,\ldots,\xi_{n-1}) \in M^n : \xi_0 = x$$

and $A_n(\xi) \leq s\}$. *Then for any* $N,\delta,\beta > 0$ *there is* $\epsilon_0 > 0$ *such that if* $0 \leq \epsilon \leq \epsilon_0$, $s \geq 0$ *and* $1 \leq n \leq N$ *one has*

$$P_x^\epsilon\{\mathrm{dist}_n(X^\epsilon,\Phi_x(n,s)) \geq \delta\} \leq \exp\left[-\frac{(s-\beta)}{s}\right] \qquad (5.20)$$

where $\mathrm{dist}_n(\xi,\Phi_x(n,s)) = \inf\limits_{\zeta\in\Phi(s)} \mathrm{dist}_n(\xi,\zeta)$.

Proof. Denote

$$\Xi_x(n,s,\delta) = \{\xi = (\xi_0,\ldots,\xi_{n-1}) : \mathrm{dist}_n(\xi,\Phi_x(n,s) \geq \delta\}$$

and

$$\Psi_\gamma(\xi) = \{\zeta = (\zeta_0,\ldots,\zeta_{n-1}) : \mathrm{dist}_n(\xi,\zeta) < \gamma\}.$$

Since M is compact now then M^n is also compact and for each $\gamma > 0$ we have a finite cover of M^n by balls $\Psi_\gamma(\xi_i)$. Take $\gamma < \delta$ then if $\psi_\gamma(\xi) \cap \Xi_x(n,s,\delta) \neq \phi$ then

$\psi_\gamma(\xi) \cap \Phi_x(n,s) = \phi$, and so $A_n(\xi) > s$. Choose $\gamma > 0$ even smaller to achieve that for such ξ,

$$P_x^\epsilon\{\mathrm{dist}_n(X^\epsilon,\xi) < \gamma\} \le \exp\left[-\frac{(s-\frac{1}{2}\beta)}{\epsilon}\right]. \tag{5.21}$$

Now from a finite cover of M^n by balls $\psi_\gamma(\xi_i)$ take the balls $\psi_\gamma(\xi_i')$ satisfying $\psi_\gamma(\xi_i') \cap \Xi_x(n,s,\delta) \ne \phi$. Then we obtain by (5.21),

$$P_x^\epsilon\{\mathrm{dist}_n(X^\epsilon,\Phi_x(n,s)) \ge \delta\} \tag{5.22}$$

$$\le \sum_{i=1}^{N_\gamma} P_x^\epsilon\{\mathrm{dist}_n(X^\epsilon,\xi_i') < \gamma\}$$

$$\le N_\gamma \exp\left[-\frac{(s-\frac{1}{2}\beta)}{\epsilon}\right]$$

where N_γ is the minimal number of balls $\psi_\gamma(\xi)$ covering M^n. For $\epsilon > 0$ small enough (5.22) implies (5.20). □

To discuss the problems concerning the asymptotical behavior of invariant measures of Markov chains X_n^ϵ we shall need the following general result on invariant measures of induced Markov chains.

Proposition 5.3. *Let* X_n *be a Markov chain in a metric space* M *with transition probabilities* $P(x,\Gamma)$ *having an invariant measure* $\mu \in \mathcal{P}(M)$. *For a Borel set* $V \subset M$ *with* $\mu(V) > 0$ *define another Markov chain* ${}^V X_n$ *(called the induced Markov chain) on* V *by its transition probabilities* ${}^V P(x,\Gamma)$ *having the form*

$${}^V P(x,\Gamma) = P_x\{X_{\tau_V} \in \Gamma\} \tag{5.23}$$

where $\tau_V = \inf\{n > 0 : X_n \in V\}$, Γ *is a Borel subset of* V *and* $P_x\{\ \}$ *denotes the probability for the Markov chain* X_n *starting at* x. *Then the restriction* $\mu_V \in \mathcal{P}(V)$ *of* $(\mu(V))^{-1}\mu$ *to* V *is the invariant measure of the Markov chain* ${}^VX_n$.

Proof. For a Borel set $\Gamma \subset V$ we have

$$ {}^VP(x,\Gamma) = \sum_{n=1}^{\infty} P_x\{\Omega_n(\Gamma)\} \tag{5.24}$$

where the events $\Omega_n(\Gamma) = \{X_n \in \Gamma \text{ and } X_k \in M\backslash V \text{ for all } k = 1,\dots,n-1\}$ are, clearly, disjoint for $n = 1,2,\dots$. Thus $1 \geq {}^VP(x,V) \geq {}^VP(x,\Gamma)$, and so for each $x \in M$,

$$\sum_{n=N}^{\infty} P_x\{\Omega_n(\Gamma)\} \to 0 \quad \text{as} \quad N \to \infty \tag{5.25}$$

for any Borel $\Gamma \subset V$. By the Markov property for $n > 1$,

$$P_x\{\Omega_n(\Gamma)\} = \int_{M\backslash V} P(x,dy)P_y\{\Omega_{n-1}(\Gamma)\}. \tag{5.26}$$

Since μ is invariant, i.e.,

$$\mu(\Gamma) = \int_M d\mu(x)P(x,\Gamma) \tag{5.27}$$

then by (5.26) for $n > 1$ one has

$$\int_V d\mu(x)P_x\{\Omega_n(\Gamma)\} = \int_{M\backslash V} d\mu(x)P_x\{\Omega_{n-1}(\Gamma)\} \tag{5.28}$$
$$- \int_{M\backslash V} d\mu(x)P_x\{\Omega_n(\Gamma)\}.$$

Taking into account that $P_x\{\Omega_1(\Gamma)\} = P(x,\Gamma)$ we derive from (5.24), (5.27) and (5.28) that

$$\int_V d\mu(x)^V P(x,\Gamma) = \mu(\Gamma) - \int_{M\setminus V} d\mu(x) P_x\{\Omega_N(\Gamma)\} \qquad (5.29)$$

$$+ \sum_{n=N+1}^{\infty} \int_V d\mu(x) P_x\{\Omega_n(\Gamma)\}.$$

In view of (5.25), letting $N \to \infty$ we obtain by Lebesgue's integral convergence theorem that

$$\int_V d\mu(x)^V P(x,\Gamma) = \mu(\Gamma). \qquad (5.30)$$

Dividing both parts of (5.30) by $\mu(V)$ we obtain the invariance of $\mu_V = (\mu(V))^{-1}\mu$. Since $\mu_V(V) = 1$ we see also from (5.30) that

$${}^V P(x,V) = 1 \quad \text{for } \mu_V\text{-almost all } x \in V. \qquad (5.31)$$

The proof is complete. □

In the remaining part of this section our arguments will follow the line of exposition in Wentzell and Freidlin [WF] and [FW] with simplications due to the discrete time case we are dealing with, which will enable us to obtain their result as a partial case.

Lemma 5.2. *Let* K *be a basic* ρ*-equivalence class. Then for any* $\delta > 0$ *there exists* $N_\delta > 0$ *such that whenever* $x,y \in K$ *one can find a sequence* $\xi = (\xi_0,\dots,\xi_{n-1})$ *with* $\xi_0 = x$, $\xi_{n-1} = y$, $n \le N_\delta$, $\max_{0\le i<n} \operatorname{dist}(\xi_i,K) < \delta$, *and* $A_n(\xi) < \delta$.

Proof. Since M is compact (Assumption 5.3) then by Lemma 5.1 K is a closed, and so a compact set. Choose points $x_1,\dots,x_m \in K$ such that $\bigcup_{\ell=1}^{m} U_r(x_\ell) \supset K$ where $r = \sup\{\gamma : \eta_\gamma < \frac{\delta}{3}\}$ and η_γ was defined in (5.17). Let

$x \in U_r(x_k)$ and $y \in U_r(x_\ell)$. If $\xi^{(1)} = (\xi_0^{(1)},\dots,\xi_{n-1}^{(1)})$ and $\xi^{(2)} = (\xi_0^{(2)},\dots,\xi_{n-1}^{(2)})$ are two sequences such that $\xi_i^{(1)} = \xi_i^{(2)}$ for all $i = 1,\dots,n-2$ and $\xi_0^{(1)} = x$, $\xi_{n-1}^{(1)} = y$, $\xi_0^{(2)} = x_k$, $\xi_{n-1}^{(2)} = x_\ell$ then it follows

$$|A_n(\xi^{(1)}) - A_n(\xi^{(2)})| \le \frac{2\delta}{3} . \tag{5.32}$$

Thus it suffices to show that for each couple $x_\ell, x_k \in K$ one can find a sequence $\xi^{\ell,k} = \{\xi_0^{\ell,k},\dots,\xi_{n_{\ell,k}-1}^{\ell,k}\}$ such that $A_n(\xi^{\ell,k}) < \frac{\delta}{3}$ and $\mathrm{dist}(\xi_i^{\ell,k},K) < \delta$. By the definition of the ρ-equivalence there exist sequences $\xi^{(i)} = \{\xi_0^{(i)},\dots,\xi_{n_i-1}^{(i)}\}$ such that $\xi_0^{(i)} = x_k$, $\xi_{n_i-1}^{(i)} = x_\ell$, and $A_{n_i}(\xi^{(i)}) \to 0$ as $i \to \infty$. If i is big enough all points of $\xi^{(i)}$ must belong to the δ-neighborhood of K since for otherwise we shall have a convergent sequence $\xi_{\nu_j}^{(i_j)} \to z$ with $\mathrm{dist}(z,K) \ge \delta$. On the other hand, clearly $V(x_k,z) = V(z,x_\ell) = 0$, and so z is ρ-equivalent to all points of K which is the contradiction. □

Lemma 5.3. *Let* K *be a compact subset of* M *which does not contain entirely any orbit* $\{F^\ell x, \ell \ge 0\}$, $x \in K$ *of the map* F. *Then there exist numbers* $a = a(K) > 0$ *and* $N = N(K) > 0$ *such that*

(a) *for any sequence* $\xi = (\xi_0,\dots,\xi_{n-1})$ *with* $n > N$ *and* $\xi_i \in K$, $i = 0,\dots,n-1$ *one has* $A_n(\xi) > (n-N)a$;

(b) *there exists* $\epsilon_0 > 0$ *such that for any* $n > N$,

$$P_x^\epsilon\{\tau_{M\setminus K} > n\} \le e^{-\frac{(n-N)a}{\epsilon}} \tag{5.33}$$

provided $x \in K$ *and* $0 < \epsilon < \epsilon_0$, *where* $\tau_V = \inf\{m > 0 : X_m^\epsilon \in V\}$.

Proof. For $x \in K$ put $\nu(x) = \inf\{\ell : F^\ell x \notin K\}$. By the assumption $\nu(x) < \infty$ for any $x \in K$. In the same way as in the proof of Theorem 3.1 we see that the function ν is upper semicontinuous, and so it achieves its maximum $\tilde{N} = \max_{x \in K} \nu(x) < \infty$ on K. Put $N_1 = \tilde{N}+1$. Consider the function A_{N_1} on sequences staying entirely in K. Then A_{N_1} is a continuous function on N_1-fold product K^{N_1} which is a compact set. Since K does not contain an orbit of F of length N_1 then $b_1 = \min_{\xi \in K^{N_1}} A_{N_1}(\xi) > 0$. From the additive structure of functions A_n it follows that for $n > N_1$,

$$A_n(\xi) \geq b[\frac{n}{N_1}] > \frac{b_1}{N_1}(n-N_1) \tag{5.34}$$

for any sequence $\xi = (\xi_0, \ldots, \xi_{n-1})$ such that $\xi_i \in K$ for all $i = 0, 1, \ldots, n-1$, where $[\cdot]$ denotes the integral part of a number. This proves the assertion (a) with $N = N_1$ and $a = \frac{b}{N_1}$.

Denote $K_\delta = \{y : \mathrm{dist}(y, K) \leq \delta\}$. Clearly, K_δ is a compact set and if $\delta > 0$ is small enough then K_δ does not contain an entire orbit of F, and so we can repeat arguments of the item (a) for K_δ in place of K. This gives that any sequence $\xi = (\xi_0, \ldots, \xi_{n-1})$ staying entirely in K_δ satisfies $A_n(\xi) > b_2(n-N_2)$ for some $b_2, N_2 > 0$ independent of ξ. For $x \in K$ put $\Phi_x(n) = \{\xi = (\xi_0, \ldots, \xi_{n-1}) : \xi_0 = x \text{ and } A_n(\xi) \leq b_2(n-N_2)\}$. Then for $n > N_2$ no sequence from $\Phi_x(n)$ stays entirely in K_δ. Thus

$$\{\tau_{M\setminus K} > n\} = \{X_i^\epsilon \in K \text{ for all } i = 0,1,\ldots,n\}$$

$$\subset \{\mathrm{dist}_n(X^\epsilon, \Phi_x(n)) \geq \delta\}.$$

Take here $n = n_0 = N_2+1$ then by Corollary 5.2

$$P_x^\epsilon\{\tau_{M\setminus K} > n_0\} \leq P_x^\epsilon\{\mathrm{dist}_{n_0}(X^\epsilon, \Phi_x(n_0)) \geq \delta\} \leq e^{-\frac{b_2}{2\epsilon}}$$

provided $\epsilon > 0$ is small enough. Then by the Markov property (cf. (3.11)) for any $n \geq n_0$,

$$P_x^\epsilon\{\tau_{M\setminus K} > n\} \leq P_x^\epsilon\{\tau_{M\setminus K} > n_0[\tfrac{n}{n_0}]\} \leq \exp\left[-\frac{b_2[\frac{n}{n_0}]}{2\epsilon}\right]$$

$$\leq \exp\left[-\frac{b_2}{2n_0\epsilon}(n-n_0)\right]$$

proving (5.33) with $a = \frac{b_2}{2n_0}$ and $N = n_0$. □

In the remaining part of this section we shall need also the following

Assumption 5.4. There exists only a finite number of basic ρ-equivalence classes $K_1,\ldots,K_\nu$.

By Lemma 5.1 and Assumption 5.3, $K_1,\ldots,K_\nu$ are compacts. By Proposition 5.1 ρ-quasiattractors among the compacts K_i (and, clearly, only them) are also attractors. Let V_i be open sets such that

$$K_i \subset V_i \subset U_r(K_i) = \{y : \mathrm{dist}(y, K_i) < r\}. \qquad (5.35)$$

We shall always take $r > 0$ small enough so that V_i, $i = 1,\dots,\nu$ are disjoint. Denote $V = \bigcup_{1\le i\le \nu} V_i$ and consider the Markov chain ${}^V X_n^\epsilon$ introduced in the same way as in Proposition 5.3 by means of transition probabilities ${}^V P^\epsilon(x,\Gamma) = P_x\{X^\epsilon_{\tau_V} \in \Gamma\}$ where $\tau_V = \inf\{n > 0 : X_n^\epsilon \in V\}$ and Γ is a Borel subset of V. Since K_i and K_j are equivalence classes the value $B(x,y)$ defined by (5.3) remains the same for all $x \in K_i$ and $y \in K_j$, and it will be denoted B_{ij}. Clearly, if $i \ne j$ at least one of the numbers B_{ij} and B_{ji} is positive. It is clear from the definition that K_i is a ρ-quasiattractor (and so it is an attractor) if and only if $B_{ij} > 0$ for any $j \ne i$.

In the following result we estimate the asymptotics of transition probabilities of the Markov chain ${}^V X_n^\epsilon$.

Lemma 5.4. *Given* $\beta > 0$ *one can find a number* $r = r(\beta) > 0$ *and an integer* $N = N_0(\beta) > 0$ *such that for any choice of open sets* $V_i \subset U_r(K_i)$, $i = 1,\dots,\nu$ *there exists* $\epsilon_0 > 0$ *so that the N-step transition probabilities of the Markov chain* ${}^V X_n^\epsilon$ *satisfy*

$$\exp(-(B_{k\ell}+\beta)/\epsilon) > {}^V P^\epsilon(N,x,V_\ell) < \exp((-B_{k\ell}+\beta)/\epsilon) \qquad (5.36)$$

provided $x \in V_k$, $0 < \epsilon < \epsilon_0$, *and* $1 \le k,\ell \le \nu$.

Proof. By the Chapman-Kolmogorov formula we have for a fixed integer $n > 0$,

$$
\begin{aligned}
&{}^{V}P^{\epsilon}(n,x,V_\ell) \qquad (5.37)\\
&= \int_V \cdots \int_V {}^{V}P^{\epsilon}(x,dz_1)\,{}^{V}P^{\epsilon}(z_1,dz_2)\\
&\qquad \times \cdots \times {}^{V}P^{\epsilon}(z_{n-2},dz_{n-1})\,{}^{V}P^{\epsilon}(z_{n-1},V_\ell)\\
&= \sum_{i_1,\dots,i_{n-1}} \int_{V_{i_1}} \cdots \int_{V_{i_{n-1}}} \int_{V_\ell} {}^{V}P^{\epsilon}(x,dz_1)\,{}^{V}P^{\epsilon}(z_1,dz_2)\\
&\qquad \times \cdots \times {}^{V}P^{\epsilon}(z_{n-1},z_n)\\
&\le \sum_{i_1,\dots,i_{n-1}} \sup_{z\in V_k} {}^{V}P^{\epsilon}(z,V_{i_1}) \sup_{z\in V_{i_1}} {}^{V}P^{\epsilon}(z,V_{i_2})\\
&\qquad \times \cdots \times \sup_{z\in V_{i_{n-2}}} {}^{V}P^{\epsilon}(z,V_{i_{n-1}}) \sup_{z\in V_{i_{n-1}}} {}^{V}P^{\epsilon}(z,V_\ell).
\end{aligned}
$$

Clearly, $B_{k\ell} \le B_{ki_1} + B_{i_1i_2} + \dots + B_{i_{n-2}i_{n-1}} + B_{i_{n-1}\ell}$, and so in order to prove the right hand side of (5.36) it suffices to show that for all $k,\ell = 1,\dots,\nu$,

$$\sup_{z\in U_k} {}^{V}P^{\epsilon}(z,U_\ell) < \exp((-B_{k\ell} + \tfrac{\beta}{n})/\epsilon). \qquad (5.38)$$

If $k = \ell$ there is nothing to prove since $B_{kk} = 0$. Let $k \ne \ell$. Choose $r > 0$ (appearing in the statement of this lemma) and another number $\delta > 0$ so small that $\eta_{r+\delta} < \frac{\beta}{3n}$ and $\overline{U_\delta(K_\ell)} \subset U_\ell$ for each ℓ, where η_γ is defined by (5.17). Then any sequence $\xi = (\xi_0,\dots,\xi_{m-1})$ with $\xi_0 \in V_k$ and $\xi_{m-1} \in U_{r+\delta}(K_\ell)$ must satisfy:

$$A_m(\xi) > A_m(\tilde{\xi}) - \frac{2\beta}{3n} \ge B_{k\ell} - \frac{2\beta}{3n} \qquad (5.39)$$

where $\tilde{\xi}$ is another sequence coinciding with ξ except for the first and the last points which belong to K_k and K_ℓ, respectively. Remark that $M\backslash V = M\backslash \bigcup_{1\leq i\leq \nu} V_i$ cannot contain an entire orbit $\{F^j y, j \geq 0\}$ since for otherwise $M\backslash V$ would not be disjoint from basic ρ-equivalence classes. Thus we can employ Lemma 5.3 deriving that there exists $N_1 = N_1(V)$ such that

$$P_z^\epsilon\{\tau_V > N_1\} \leq \exp\left[-\frac{B_{ij}}{\epsilon}\right] \tag{5.40}$$

for any $i,j = 1,\ldots,\nu$ and $z \in M\backslash V$ provided $\epsilon > 0$ is small enough. Any path of the Markov chain X_n^ϵ starting at $x \in V_k$ and getting to V_ℓ at the moment τ_V has only two alternatives: either $X_i^\epsilon \in M\backslash V$ for all $i = 1,\ldots N_1$ or $X_j^\epsilon \in V_\ell$ for some $j \geq 1$ and $j \leq N_1$. In the second case, in view of (5.39), $\mathrm{dist}_{N_1+1}(X^\epsilon, \Phi_x) \geq \delta$ where

$$\Phi_x = \{\xi = (\xi_0,\ldots,\xi_{N_1}) : \xi_0 = x, A_{N_1+1}(\xi) \leq B_{k\ell} - \frac{2\beta}{3n}\}.$$

Thus by Corollary 5.2 and by (5.40) for any $z \in U_k$,

$$^V P^\epsilon(z,U_\ell) \leq \int_{M\backslash V} P^\epsilon(z,dy)P_y^\epsilon\{\tau_V > N_1\} \tag{5.41}$$

$$+ P_z^\epsilon\{\mathrm{dist}_{N_1+1}(X^\epsilon,\Phi_x) \geq \delta\}$$

$$\leq \exp\left[-\frac{B_{k\ell}}{\epsilon}\right] + \exp((-B_{k\ell} + \frac{5\beta}{6n})/\epsilon)$$

provided $\epsilon > 0$ is small enough. This proves (5.38), and so it gives the upper bound in (5.36) which holds true for any $N = 1,2,3,\ldots$ provided $\epsilon_0 = \epsilon_0(N) > 0$ is small enough.

Next we shall prove the left hand side of (5.36) where we shall have to make a particular choice of $N = N_0(\beta)$. In this part of the proof we take $r > 0$ and $\delta > 0$ such that $\delta < \frac{\beta}{6\nu}$, $\eta_{r+\delta} < \frac{\beta}{6\nu}$ and $U_{2\delta}(K_i) \subset V_i$ for all $i = 1,\dots,\nu$. By the definition there exists a sequence $\xi = (\xi_0,\dots,\xi_{n-1})$ with $\xi_0 \in K_k$, $\xi_{n-1} \in K_\ell$ and $A_n(\xi) < B_{k\ell} + \frac{\beta}{6}$. This sequence may have some points belonging to the sets $U_{r+\delta}(K_i)$ for $i \neq k,\ell$. Throwing away some of the points from ξ we can construct a subsequence $\tilde{\xi}$ consisting of pieces $\{\xi_i, i_\kappa \le i \le j_\kappa\}$, $\kappa = 1,\dots,m$ of the sequence ξ such that $\xi_{i_\kappa} \in U_{r+\delta}(K_{\ell_{\kappa-1}})$, $\xi_{j_\kappa} \in U_{r+\delta}(K_{\ell_\kappa})$, $\xi_i \notin \bigcup_{1\le j\le\nu} U_{r+\delta}(K_j)$ for all $i = i_\kappa+1,\dots,j_\kappa-1$, $\kappa = 1,\dots,m$, where we put $\ell_0 = k$ and $\ell_m = \ell$. All ℓ_κ are supposed to be different here, and so $m \le \nu-1$. Remark that

$$\sum_{1\le\kappa\le m} \sum_{i_\kappa\le i\le j_\kappa-1} \rho(\xi_i,\xi_{i+1}) < B_{k\ell} + \frac{\beta}{6}. \tag{5.42}$$

Next, we shall construct another sequence $\zeta = (\zeta_0,\dots,\zeta_{\tilde{n}-1})$ which starts at $x \in V_k$, ends in K_ℓ, satisfies $A_{\tilde{n}}(\zeta) < B_{k\ell} + \frac{2}{3}\beta$, and such that among $\zeta_1,\dots,\zeta_{\tilde{n}-1}$ exactly $N = \nu N_\delta$ points belong to $\bigcup_{1\le j\le\nu} U_\delta(K_j)$ but all other points do not belong to $\bigcup_{1\le j\le\nu} U_{r+\delta}(K_j)$, where N_δ was determined in Lemma 5.2 and $\delta > 0$ was chosen above. We shall proceed in the following way. For $\kappa = 1,\dots,m-1$ let $z^{(\kappa)}, \tilde{z}^{(\kappa)} \in K_{\ell_\kappa}$ be closest to ξ_{j_κ} and $\xi_{i_{\kappa+1}}$, respectively, points of K_{ℓ_κ}. Then by the choice of $r > 0$ and $\delta > 0$, since $\xi_{j_\kappa}, \xi_{i_{\kappa+1}} \in U_{r+\delta}(K_{\ell_\kappa})$, we obtain

$$|\rho(\xi_{j_\kappa-1},\xi_{j_\kappa}) - \rho(\xi_{j_\kappa-1},z^{(\kappa)})| \qquad (5.43)$$

$$+ |\rho(\xi_{i_{\kappa+1}},\xi_{i_{\kappa+1}+1}) - \rho(\tilde{z}^{(\kappa)},\xi_{i_{\kappa+1}+1})| \le \frac{\beta}{3\upsilon} .$$

By Lemma 5.2 there exists a sequence $\eta^{(\kappa)} = (\eta_0^{(\kappa)},\ldots,\eta_{n_\kappa-1}^{(\kappa)})$ such that $n_\kappa \le N_\delta$, $\eta_0^{(\kappa)} = z^{(\kappa)}$, $\eta_{n_\kappa-1}^{(\kappa)} = \tilde{z}^{(\kappa)}$, $\eta_i^{(\kappa)} \in U_\delta(K_{\ell_\kappa})$ for all $i = 0,\ldots,n_\kappa-1$, and $A_{n_\kappa}(\eta^{(\kappa)}) < \delta < \frac{\beta}{6\upsilon}$. We shall take an arbitrary point $\xi_{i_{m+1}} \in K_\ell$ and use the same construction to obtain the sequence $\eta^{(m)}$. We shall construct also $\eta^{(0)}$ in the same way as above by taking $\xi_{j_0} = z^{(0)} = Fy^{(0)} \in K_k$ where $y^{(0)} \in K_k$ is the closest to x point in K_k. Clearly $\rho(x,z^{(0)}) < \beta/6\upsilon$, Let $\tilde{\xi}^{(\kappa)} = \{\xi_i, i_\kappa < i < j_\kappa\}$, $\kappa = 1,\ldots,m$. Consider the sequence $\tilde{\zeta}$ built by sticking together the following pieces

$$\tilde{\zeta} = (\eta^{(0)},\tilde{\xi}^{(1)},\eta^{(1)},\ldots,\tilde{\xi}^{(m)},\eta^{(m)}).$$

Let $\tilde{N}$ be the number of points of $\tilde{\zeta}$ which belong to $\bigcup_{1\le i\le\upsilon} U_\delta(K_i)$. Then $\tilde{N} \le N = \upsilon N_\delta$. If $\tilde{N} = N$ then we put $\zeta = (x,\tilde{\zeta})$. If $\tilde{N} < N$ then we define the sequence ζ as the sequence $(x,\tilde{\zeta})$ followed by the $(n-\tilde{N})$ points:

$$F\xi_{i_{m+1}},\ F^2\xi_{i_{m+1}},\ldots,F^{N-\tilde{N}}\xi_{i_{m+1}} \in K_\ell .$$

These last points do not add anything to the sum $\sum_i \rho(\zeta_i,\zeta_{i+1})$, and so in view of (5.42), (5.43), and the inequalities $A_{n_\kappa}(\eta^{(\kappa)}) < \frac{\beta}{6\upsilon}$ and $\rho(x,z^{(0)}) < \frac{\beta}{6\upsilon}$, we obtain that $A_{\tilde{n}}(\zeta) < B_{k\ell} + \frac{2}{3}\beta$, where $\tilde{n}$ is the number of points in ζ. Since exactly N points among $\zeta_1,\ldots,\zeta_{\tilde{n}-1}$ belong to $\bigcup_{1\le i\le v} U_\delta(K_i)$ (we disregard $\zeta_0 = x$ here) all other points do not belong to $\bigcup_{1\le i\le v} U_{r+\delta}(K_i)$, and $U_{2\delta}(K_i) \subset V_i \subset U_r(K_i)$ for all $i = 1,\ldots,v$ then

$$\{{}^V X_N^\epsilon \in V_\ell \text{ provided } {}^V X_0^\epsilon = x\} \supset \{\text{dist}_{\tilde{n}}(X^\epsilon,\zeta) < \delta\},$$

and so

$${}^V P^\epsilon(N,x,V_\ell) > P_x^\epsilon\{\text{dist}_{\tilde{n}}(X^\epsilon,\zeta) < \delta\}.$$

Employing Theorem 5.2 (a) we derive the left hand side of (5.36). □

Next, we shall need a couple of simple results from Wentzell and Freidlin [WF] concerning Markov chains. Let L be a finite set, whose elements will be denoted by letters i, j, k, m, n, etc. Given $i \in L$, a graph consisting of arrows $m \to n$ $(m \ne i,\ m,n \in L,\ n \ne m)$ is called an i-graph if it satisfies the following conditions: every point $m \ne i$ is the origin of exactly one arrow, and the graph has no cycles. Remark that the last condition can be replaced by the following one: for any point $m \ne i$ there exists a sequence of arrows leading from it to the point i. The set of all i-graphs will be denoted by $G(i)$.

Lemma 5.5. *Consider a Markov chain with set of states* L *and transition probabilities* p_{ij} *and assume that every state can be reached from any other state in a finite number of steps. Then the stationary distribution of the chain is* $\{\theta_i(\sum_{i\in L} \theta_i)^{-1}, i \in L\}$, *where*

$$\theta_i = \sum_{g\in G(i)} \theta(g), \theta(g) = \prod_{(m\to n)\in g} p_{mn}, \tag{5.44}$$

and Π *denotes the product.*

Proof. The numbers θ_i are positive. It suffices to show that they satisfy the system of equations

$$\theta_i = \sum_{j=L} \theta_j p_{ji}, \text{ for all } i \in L, \tag{5.45}$$

since our assumption says that some power of the matrix (p_{ij}) is a positive matrix, and so by the well known finite Markov chains counterpart of Proposition 1.8 (see Doob [Do], Ch.V,§2) it follows that the stationary distribution is the unique (up to a multiplicative constant) solution of this system. Writing $\theta_i = \theta_i \sum_{j\in L} p_{ij}$ we see that (5.45) is equivalent to

$$\theta_i \sum_{k\neq i} p_{ik} = \sum_{j\neq i} \theta_j p_{ji}. \tag{5.46}$$

If we substitute in (5.46) the numbers θ_i defined by (5.44), then on both sides of (5.46) we obtain the sum of $\theta(g)$ over all graphs g satisfying the following conditions: every point $m \in L$ originates exactly one arrow $m \to n$, $n \neq m$, $n \in L$, and g has exactly one cycle which contains i. Hence (5.46) turns out to be an equality. □

Lemma 5.6. *Let a Markov chain be given on an arbitrary state space* V *divided into disjoint subsets* V_i, *where* i *runs over a finite set* L. *Suppose that for some positive numbers* p_{ij}, $j \neq i$, $i,j \in L$, *a number* $a > 1$, *and an integer* $N > 0$ *the N-step transition probabilities* $P(N,x,\Gamma)$, $\Gamma \subset V$ *of the Markov chain satisfy*

$$a^{-1}p_{ij} \leq P(N,x,V_j) \leq ap_{ij},\ x \in V_i,\ i \neq j. \tag{5.47}$$

Then for any invariant measure $\mu \in \mathcal{P}(V)$ *of this Markov chain one has*

$$a^{2-2\nu} \frac{\theta_i}{\theta_1+\ldots+\theta_\nu} \leq \mu(V_i) \leq a^{2\nu-2} \frac{\theta_i}{\theta_1+\ldots+\theta_\nu}, \tag{5.48}$$

where ν *is the number of elements in* L *and* θ_i *are defined in Lemma 5.5.*

Proof. Since

$$\int_V d\mu(x)P(N,x,V_i) = \mu(V_i)$$

and all $P(N,x,V_i) > 0$ then $\mu(V_i) > 0$, $i = 1,\ldots,\nu$. Consider a finite Markov chain on L having transition probabilities

$$\tilde{p}_{ij} = \frac{1}{\mu(V_i)} \int_{V_i} \mu(dx)P(N,x,V_j). \tag{5.49}$$

The stationary distribution of this Markov chain is $\{\mu(V_i), i \in L\}$ which can be expressed by means of the formula of Lemma 5.5 with $\tilde{p}_{ij}$ in place of p_{ij}. Now taking into account (5.47) and (5.49) we derive (5.48). □

Let $L = \{1,\ldots,\nu\}$, $i \in L$,

$$B(i) = \min_{g \in G(i)} \sum_{(m\to n)\in g} B_{mn}$$

and

$$L_{min} = \{i \in L : B(i) = \min_{j \in L} B(j)\}.$$

Now we can formulate the main result of this section.

Theorem 5.4. *If* $i \in L_{min}$ *then* K_i *is a* ρ*-quasiattractor. Let* $\Gamma \subset M$ *be a closed set disjoint with* $\bigcup_{i \in L_{min}} K_i$. *Then any invariant measure* $\mu^\epsilon \in \mathcal{P}(M)$ *of the Markov chain* X_n^ϵ *satisfy*

$$\lim_{\epsilon \to 0} \mu^\epsilon(\Gamma) = 0. \tag{5.50}$$

Proof. It is clear from the definition of our partial order that if K_i, $i \in L$ is not a ρ-quasiattractor then there exists a ρ-quasiattractor K_j, $j \neq i$, $j \in L$ such that $B_{ij} = 0$. Let g be an arbitrary i-graph. Since $j \neq i$ then there is a unique arrow $j \to k$ originating at j and leading to some $k \in L$. Cancel this arrow and add the new arrow $i \to j$. We shall obtain then a j-graph $\tilde{g}$. Since $B_{jk} > 0$ and $B_{ij} = 0$ then

$$\sum_{(m\to n)\in g} B_{mn} > \sum_{(m\to n)\in \tilde{g}} B_{mn},$$

and so the above sums cannot attain the minimum on i-graphs, i.e., $i \notin L_{min}$.

To prove (5.50) we shall choose disjoint open neighborhoods $V_i \subset U_r(K_i)$ of K_i with $r > 0$ small enough. Employing Proposition 5.3 we conclude that for any invariant measure $\mu^\epsilon \in \mathcal{P}(M)$ of the Markov chain X_n^ϵ the measure $(\mu^\epsilon(V))^{-1}\mu^\epsilon$ on $V = \bigcup_{i \in L} V_i$ is invariant with respect to the Markov chain ${}^V X_n^\epsilon$ provided $\mu^\epsilon(V) > 0$. This last condition holds true since in view of Lemma 5.3 (b), if $n > 0$ is a big enough integer then

$$P_x^\epsilon\{\tau_V > n\} < \frac{1}{2}$$

for all $x \in M \setminus V$ and $\epsilon > 0$ small enough. Then $\sum_{k=1}^{n} P^{\epsilon}(k,x,V) > \frac{1}{2}$, for $x \in M\setminus V$, and so

$$n\mu^{\epsilon}(V) = \int_M d\mu^{\epsilon}(x) \sum_{k=1}^{n} P^{\epsilon}(k,x,V) \geq \frac{1}{2}\mu^{\epsilon}(M\setminus V).$$

Thus $\mu^{\epsilon}(V) > (n+\frac{1}{2})^{-1} > 0$.

Next, in order to estimate invariant measures of the Markov chain ${}^{V}X_n^{\epsilon}$ we can apply the machinery developed in Lemmas 5.4-5.6. We choose $r > 0$ so small that $\bigcup_{i\in L_{min}} U_r(K_i)$ is disjoint with the closed set Γ, and such that the inequality (5.36) holds true with $\beta > 0$ satisfying

$$\beta < (2\nu)^{-1}\left[\min_{j\notin L_{min}} B(j) - \min_{i\in L} B(i)\right].$$

Then by Lemma 5.6 we derive that

$$(\mu^{\epsilon}(V))^{-1}\mu^{\epsilon}\left(\bigcup_{j\notin L_{min}} V_j\right) < e^{-\frac{\gamma}{\epsilon}} \tag{5.51}$$

for some $\gamma > 0$ and $\epsilon > 0$ small enough, and so

$$(\mu^{\epsilon}(V))^{-1}\mu^{\epsilon}\left[\bigcup_{i\in L_{min}} V_i\right] \to 1 \quad \text{as} \quad \epsilon \to 0. \tag{5.52}$$

It remains to show that

$$\mu^{\epsilon}(M\setminus V) \to 0 \quad \text{as} \quad \epsilon \to 0 \tag{5.53}$$

(for more precise estimates see Remark 5.2 below). Let $\mu^{\epsilon_i} \xrightarrow{w} \mu$ then by Theorem 1.1 μ is an F-invariant measure. From Lemma 5.3 (a) it follows that $M\setminus V$ is a subset of the set of wandering points for F and so by Corollary 4.2, $\mu(M\setminus V) = 0$. Since $M\setminus V$ is closed then by the basic result about the weak convergence (see Theorem 2.1 in Billingsley [Bi]) it follows that

$$\limsup_{i\to\infty} \mu^{\epsilon_i}(M\setminus V) \le \mu(M\setminus V) = 0.$$

This being true for any subsequence $\epsilon_i \to 0$, yields (5.53) and completes the proof. □

Remark 5.2. One can refine both (5.51) and (5.52). Since $\mu^{\epsilon}(V) \to 1$ as $\epsilon \to 0$ then the same arguments as in the proof of Theorem 5.4 lead to the estimate

$$\exp\{-\epsilon^{-1}(B(j)-\min_{j\in L} B(i)+\beta)\} < \mu^{\epsilon}(V_j) \tag{5.54}$$

$$< \exp\{-\epsilon^{-1}(B(j)-\min_{i\in L} B(i)-\beta)\}$$

where $\beta > 0$ can be chosen arbitrarily small provided $0 < \epsilon < \epsilon(\beta)$ is sufficiently small. Concerning points $x \in M\setminus \bigcup_{i\in L} K_i$ we can estimate the measure $\mu^{\epsilon}(U_\delta(x))$ of their small neighborhoods $U_\delta(x)$ in the same way as above by introducing another compact $K_{\upsilon+1} = x$. Choosing small disjoint neighborhoods of these compacts we shall obtain by the same arguments as above that

$$\exp\{-\epsilon^{-1}(\min_{1\le i\le \upsilon} (B(i)+B_i(x))-\min_{1\le i\le \upsilon} B(i)-\beta)\} \tag{5.55}$$

$$> \mu^{\epsilon}(U_\delta(x))$$

$$> \exp\{-\epsilon^{-1}(\min_{1\le i\le \upsilon} (B(i)+B_i(x))-\min_{1\le i\le \upsilon} B(i)+\beta)\}$$

where $B_i(x) = B(y,x)$ for $y \in K_i$, $1 \leq i \leq \nu$. Again $\beta > 0$ can be made arbitrarily small. In particular,

$$\mu^\epsilon(M\backslash V) < e^{-\frac{\gamma}{\epsilon}} \tag{5.56}$$

for some $\gamma > 0$ provided $\epsilon > 0$ is sufficiently small.

Corollary 5.3. *Any weak limit point* μ *of invariant measures* μ^ϵ *of* X_n^ϵ *as* $\epsilon \to 0$ *has support in* $\bigcup_{i \in L_{min}} K_i$ *which is a subset of the union of ρ-quasiattractors which, in turn, is a subset of the union of attractors. If* L_{min} *contains only one element* i_0 *and there exists a unique F-invariant measure* μ *on* K_{i_0} *then* $\mu^\epsilon \xrightarrow{w} \mu$ *as* $\epsilon \to 0$.

Proof follows immediately from Theorems 1.1 and 5.4.

Next, we shall derive from the above results Wentzel and Freidlin's [WF] theorem which deals with the asymptotic behavior of invariant measures of diffusion type random perturbations (we shall consider certain problems for this type of random perturbations in Chapter III). This model considered on a smooth Riemannian manifold M leads to a diffusion Markov process X_t^ϵ generated by operators $L^\epsilon = \epsilon L + b$, where L is an elliptic second order differential operator and b is a vector field. This means that transition probabilities $P^\epsilon(t,x,\Gamma)$ satisfy the parabolic equation $\frac{\partial P^\epsilon}{\partial t} = L^\epsilon P^\epsilon$ with the initial condition $P^\epsilon\big|_{t=0} = \chi_\Gamma$. The Markov processes X_t^ϵ are viewed as random perturbations of a flow F^t solving the ordinary differential equation

$$\frac{dF^t x}{dt} = b(F^t x),\ F^0 x = x.$$

We shall not discuss at this point specific features of such random perturbations since the only fact needed here is the following property of transition probabilities similar to (5.1):

$$\lim_{\epsilon\to 0} \epsilon \log P^{\epsilon}(t,x,U) = -\inf_{y\in U} B_t(x,y) \tag{5.57}$$

for any $x \in M$ and an open set U, where

$$B_t(x,y) = \inf_{\varphi_0=x,\varphi_t=y} A_t(\varphi) \tag{5.58}$$

$$= \inf_{\varphi_0=x,\varphi_t=y} \int_0^t \|b(\varphi_s)-\dot{\varphi}_s\|^2 ds$$

where the infimum is taken over absolutely continuous curves φ_s, $0 \leq s \leq t$ on M starting at x and ending at y, $\dot{\varphi}_s = \frac{d\varphi_s}{ds}$ denotes the tangent (speed) vector to φ_s, and $\|\cdot\|$ denotes certain Riemannian norm in the tangent bundle constructed by means of diffusion coefficients of X_t^{ϵ}. The relation (5.57) follows from more general results which can be found in Wentzell and Freidlin [WF] and [FW] (see, especially, Theorem 1.2 in Ch. 4 of [FW]), as well as in Friedman [Fri] (see, especially, Ch. 14, Theorem 6.1).

If we shall apply our theory to $F = F^1$ and X_t^{ϵ} considered only for integer $t = 0,1,2,...$ then the results concerning invariant measures will remain valid for the continuous time process X_t^{ϵ} since the invariant measures of X_t^{ϵ} will be, of course, invariant with respect to X_n^{ϵ}. The only fact needed to be checked is the coincidence of Assumption 5.4 with the corresponding assumption formulated by Wentzel and Freidlin for the continuous time case. They called x and y equivalent (written $x \sim y$) if and only if $\inf_{t\geq 0} B_t(x,y) = \inf_{t\geq 0} B_t(y,x) = 0$. Our definition of the equivalence relation which we shall denote here by $\overset{1}{\sim}$ corresponds to the case when the above infimum is taken only over integers: $x \overset{1}{\sim} y$ if and only if $\inf_{\text{integer } n\geq 0} B_n(x,y) = \inf_{\text{integer } n\geq 0} B_n(y,x) = 0$. Denote the equivalence classes containing a point x and corresponding to $\sim$ and $\overset{1}{\sim}$ by $[x]$ and $[x]^{(1)}$, respectively.

Proposition 5.5. *For any* $x \in M$, $[x] = [x]^{(1)}$.

Proof. Clearly, $[x]^{(1)} \subset [x]$. If $[x]$ consists of a single point x then $[x]^{(1)}$ is also the singleton x since our equivalence relations are reflexive. Suppose now that $[x]$ contains another point $y \neq x$. Then there exist a sequence of numbers $t_n \to \infty$ and a sequence of curves $\varphi_s^{(n)}$, $0 \leq s \leq t_n$, $\varphi_0^{(n)} = x$, $\varphi_{t_n}^{(n)} = y$ such that

$$A_{t_n}(\varphi^{(n)}) \to 0 \text{ as } t_n \to \infty. \tag{5.59}$$

Define new curves $\psi_s^{(n)} = \varphi^{(n)}_{st_n([t_n]+1)^{-1}}$ where $[\cdot]$ denotes the integral part. Then

$$B_{[t_n]+1}(x,y) \leq \int_0^{[t_n]+1} \|b(\psi_s^{(n)}) - \dot{\psi}_s^{(n)}\|^2 ds$$

$$= \int_0^{[t_n]+1} \|b(\varphi^{(n)}_{st_n([t_n]+1)^{-1}}) - \frac{t_n}{[t_n]+1} \dot{\varphi}^{(n)}_{st_n([t_n]+1)^{-1}}\|^2 ds$$

$$= \frac{[t_n]+1}{t_n} \int_0^{t_n} \|b(\varphi_u^{(n)}) - \frac{t_n}{[t_n]+1} \dot{\varphi}_u^{(n)}\|^2 du$$

$$\leq \frac{2t_n}{[t_n]+1} A_{t_n}(\varphi^{(n)}) + \frac{2}{t_n([t_n]+1)} \int_0^{t_n} \|b(\varphi_u^{(n)})\|^2 du$$

$$\leq 2A_{t_n}(\varphi^{(n)}) + 2t_n^{-1} \sup_x \|b(x)\|^2 \to 0$$

as $n \to \infty$ since the vector field b is continuous, M is compact, and so $\|b(x)\|$ is bounded. Thus

$$B(x,y) = \inf_{\text{integer } n \geq 0} B_n(x,y) = 0.$$

Similarly, $B(y,x) = 0$. Hence $x \overset{1}{\sim} y$, i.e., $y \in [x]^{(1)}$, proving Proposition 5.5. □

Proposition 5.5 in the case of continuous time random perturbations satisfying (5.57) and (5.58) enables one to pass to the discrete time case. Thus Theorem 5.4 yields Wentzell and Freidlin's [WF] result saying that if the number of basic equivalence classes is finite then limiting measures for diffusion type random perturbations must have support in the union of B_t-quasiattractors which by Proposition 5.1 turn out to be attractors.

Remark 5.3. Another model of random perturbations was studied by Gora [Gor]. He considers Markov chains $\tilde{X}_n^\epsilon$ with transition probabilities

$$\tilde{P}^\epsilon(x,\Gamma) = p\chi_\Gamma(Fx) + q\chi_\Gamma(H_\epsilon x)$$

where $p,q > 0$, $p+q = 1$, F and H_ϵ, $0 < \epsilon \leq \epsilon_0$ are continuous mappings of a compact metric space M, and $\mathrm{dist}(x,H_\epsilon(x)) \leq \epsilon$. This corresponds to a composition of independent random maps which are equal F and H_ϵ with probability p and q, respectively. Formally, this model does not agree with our definition of random perturbations of a map F given at the beginning of Section 1.1.

Nevertheless, it turns out that the invariant measures of $\tilde{X}_n^\epsilon$ coincide with invariant measures of the Markov chains X_n^ϵ having transition probabilities

$$P^\epsilon(x,\Gamma) = p\chi_\Gamma(Fx) + \sum_{k\geq 1} q^k \chi_\Gamma(H_\epsilon^k Fx)$$

which already agree with our definition. These random perturbations are not local and Assumption 5.1 may not be satisfied, so the results of Sections 1.4 and 1.5 are not directly applicable here. Still, it turns out that a method similar to one which was used in this section leads to the conclusion that weak limits as $\epsilon \to 0$ of invariant measures of X_n^ϵ have support in the union of attractors.

Remark 5.4. Though the situations considered in Sections 4 and 5, and in Remark 5.3 do not exhaust all cases in which weak limits as $\epsilon \to 0$ of invariant measures of random perturbations X_n^ϵ of a map F sit on attractors of F, it is not true in general. Consider, for instance, the following example.

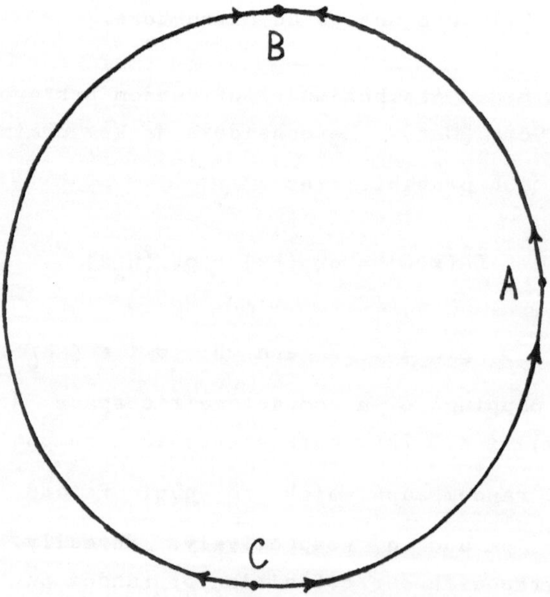

Figure 5.1

On this picture, F is a continuous map of the unit circle $M = S^1 = \{\varphi : 0 \leq \varphi \leq 2\pi\}$ having three fixed points A, B, and C with coordinates $\varphi = 0$, $\varphi = \frac{\pi}{2}$, and $\varphi = \frac{3\pi}{2}$, respectively. Arrows indicate directions in which F moves points. This can be done by taking $F = F^1$ where $\frac{dF^t x}{dt} = b(F^t x)$, b is a smooth vector field on S^1 taking on zero values only at A, B and C, and $b(\varphi)$ equals $1-\cos\varphi$, $\cos\varphi$, and $\cos\varphi$ in small neighborhoods of points A, B, and C, respectively. We shall choose random perturbations X^ϵ which perturb the map F in a neighborhood of A much weaker than in other points. For instance, the transition probabilities $P^\epsilon(x,\cdot)$ may have densities $p^\epsilon(x,y)$, i.e., $P^\epsilon(x,dy) = p^\epsilon(x,y)dy$ such that $p^\epsilon(x,y)$ is of order $\epsilon^{-1}\exp(-\epsilon^{-1}\mathrm{dist}(Fx,y))$ when x is outside of some neighborhood of A, and $p^\epsilon(x,y)$ is of order $\alpha(\epsilon)\exp(-\alpha(\epsilon)\mathrm{dist}(Fx,y))$ with $\alpha(\epsilon) = e^{\epsilon^{-3}}$ when x belongs to a neighborhood of A. Though B is the attractor and A is not, it is easy to see that Markov chains X_n^ϵ spend much more time in the $e^{-\epsilon^{-2}}$-neighborhood of A than in any neighborhood of B or C. By the ergodic theorem this will imply that for small $\epsilon > 0$ invariant measures of X_n^ϵ are "close" to the unit mass at A. Similar examples can be found in Gora's paper [Gor].

Chapter II

Random perturbations of hyperbolic and expanding transformations

In this chapter we shall study the asymptotical behavior of invariant measures, entropies, and other characteristics of random perturbations of dynamical systems with complicated dynamics satisfying certain hyperbolicity or expanding conditions.

2.1. Preliminaries.

In this section we shall discuss general ideas of our approach, as well as some assumptions we impose both on dynamical systems and their random perturbations.

Throughout this chapter $F: M \to M$ will be a C^2 endomorphism of a connected ν-dimensional locally compact Riemannian manifold M of class C^2. Both F and its random perturbations X_n^ε will satisfy certain additional conditions which we shall specify later on. In order to study invariant measures μ^ε of X_n^ε for small $\varepsilon > 0$ we shall use the formula

$$\int_M d\mu^\varepsilon(x) P^\varepsilon(n, x, \Gamma) = \mu^\varepsilon(\Gamma) \tag{1.1}$$

which by (I.1.3) and (I.1.19) holds true for any $n = 1, 2, \ldots$ and a Borel set $\Gamma \subset M$. This formula will enable us to derive asymptotical properties as $\varepsilon \to 0$ of measures μ^ε from the corresponding properties of n-step transition probabilities $P^\varepsilon(n, x, \cdot)$ of X_n^ε with suitably chosen n depending on ε. Suppose that we are in circumstances of Section 1.4 or Section 1.5 so that when $\epsilon \to 0$ the measures μ^{ϵ} will be concentrating on attractors. If there is only finite number of attractors $K_1, \ldots, K_{\ell}$ and, in addition, some kind of Corollary I.5.1 holds true then choosing disjoint open neighborhoods $V_i \supset \overline{U_r(K_i)} \supset K_i$ of K_i we can write for $n = n(\epsilon)$ and $\Gamma \subset V_{i_0}$,

$$\int_M d\mu^{\epsilon}(x) P^{\epsilon}(n,x,\Gamma) \tag{1.2}$$

$$= \sum_{1 \le i \le \ell} \int_{U_r(K_i)} d\mu^{\epsilon}(x) P^{\epsilon}(n,x,\Gamma) + o(1)$$

$$\le \int_{U_r(K_{i_0})} d\mu^{\epsilon}(x) P^{\epsilon}(n,x,\Gamma)$$

$$+ \sup_{x \in \bigcup_{i \ne i_0} U_r(K_i)} P_x^{\epsilon}\{\tau_{U_r(K_{i_0})} \le n\} + o(1)$$

$$\le \int_{U_r(K_{i_0})} d\mu^{\epsilon}(x) P^{\epsilon}(n,x,\Gamma) + n^2 e^{-\frac{\beta}{\epsilon}} + o(1),$$

for some $\beta, r > 0$, provided $\epsilon > 0$ is small enough, where $o(1) \to 0$ as $\epsilon \to 0$. This means that if n is not too big, say $n < e^{\frac{\beta}{3\epsilon}}$, then it suffices to study transition probabilities in a neighborhood of each attractor separately.

In the same way as in Corollary I.5.1 we shall see that our conditions will imply that paths of Markov chains X_n^{ϵ} which are not δ-pseudo-orbits make negligible contributions to the probability. In fact, this δ can be taken of order $\epsilon^{-(1-\alpha)}$ with $0 < \alpha < 1$. Dynamical systems we shall work with possess certain shadowing properties saying that δ-pseudo-orbits stay close to true orbits of a map F. This leads to the study of probabilities for Markov chains X_n^ϵ to stay in small tubular neighborhoods of orbits of the map F, which will be accomplished by means of a linearization procedure yielding certain Markov chains in tangent bundles. Our ultimate goal will be to show that the transition probabilities $P^\epsilon(n(\epsilon),x,\cdot)$ are absolutely continuous with respect to the Riemannian volume in the direction of certain unstable (expanding) submanifolds. Through (1.1) this property passes to μ^ϵ and, eventually, to weak limits of μ^ϵ as $\epsilon \to 0$. By some nontrivial ergodic theory's arguments one knows that for hyperbolic dynamical systems invariant measures of F are determined uniquely by the above property. This is, roughly, what we are going to do in this chapter.

Example 1.1. For the illustration of our method consider the following simple example where $F:S^1 \to S^1$ acts on the unit circle S^1 by $Fe^{ix} = e^{2ix}$, $0 \le x < 2\pi$. Assigning the coordinate x to e^{ix} we define random perturbations X_n^ϵ as Markov chains with transition probabilities

$$P^\epsilon(x,\Gamma) = \int_\Gamma \epsilon^{-1} \sum_{k=-\infty}^{\infty} q\left[\frac{2\pi k+y-2x}{\epsilon}\right]dy, \quad \Gamma \subset [0,2\pi) \qquad (1.3)$$

where $q \ge 0$ is a continuous function such that $\int_{-\infty}^{\infty} q(z)dz = 1$ and $q(z) \le \alpha^{-1}e^{-\alpha|z|}$ for some $\alpha > 0$. In particular, if $q = 0$ outside of a compact set we shall have local random perturbations as in Section 1.4. Remark, at once, that the Markov chains X_n^ϵ defined by (1.3) have for all $\epsilon > 0$ the unique invariant measure which is the Lebesgue measure denoted by mes. Hence there is no question here about the limiting measure as $\epsilon \to 0$ which will be, of course, the same mes. Still, an exhibition of our method for this example may help the reader to understand the general case. Lifting the process X_n^ϵ to the universal cover of S^1 which is the real line $\mathbb{R}^1$ $(-\infty,\infty)$ we shall obtain another Markov chain Ξ_n^ϵ on $\mathbb{R}^1$ having the representation $\Xi_n^\epsilon = \epsilon \sum_{k=1}^{n} 2^{n-k}\theta_k + 2^n\Xi_0^\epsilon$ where θ_k, $k = 1,2,\ldots$ are independent random variables with the same distribution $P\{\theta_k \in V\} = \int_V q(z)dz$, $V \subset \mathbb{R}^1$. If $P^\epsilon(n,x,\cdot)$ and $R^\epsilon(n,x,\cdot)$ are n-step transition probabilities of X_n^ϵ and Ξ_n^ϵ, respectively, then for $\Gamma \subset [0,2\pi)$,

$$P^\epsilon(n,x,\Gamma) = \sum_{k=-\infty}^{\infty} R^\epsilon(n,x,\Gamma+2\pi k) \tag{1.4}$$

$$= \sum_{k=-\infty}^{\infty} P\left\{\sum_{k=1}^{n} 2^{-k}\theta_k \in 2^{-n}\left[\frac{\Gamma+2\pi k}{\epsilon}\right] - \frac{x}{\epsilon}\right\}.$$

It is not difficult to understand (see Section 2.2) that the random variables $\sum_{k=1}^{n} 2^{-k}\theta_k$ have probability distributions with densities r_n satisfying $r_n(z) \le \beta^{-1}e^{-\beta|z|}$ where $\beta > 0$ is independent of n. For $n \ge (\log \epsilon)^2$ and $\epsilon > 0$ small enough we may assume that $2^{-n}\epsilon^{-1} < (2\pi)^{-1}$. Thus for some $C > 0$,

$$P\left\{\sum_{k=1}^{n} 2^{-k}\theta_k \in 2^{-n}\left[\frac{\Gamma+2\pi k}{\epsilon}\right] - \frac{x}{\epsilon}\right\} \tag{1.5}$$

$$\le C2^{-n}\epsilon^{-1}\beta^{-1}e^{-\beta N}\text{mes } \Gamma$$

provided $N+1 \le \epsilon^{-1}(2^{-(n-1)}\pi k-x) \le N+2$, i.e., $|k-2^{n-1}\pi^{-1}(x+\epsilon N)| \le 2^{n-1}\epsilon\pi^{-1}$. Substituting (1.5) to (1.4) we shall derive that $P^\epsilon(n,x,\Gamma) \le \tilde{C}$ mes Γ, and so by (1.1),

$\mu^\epsilon(\Gamma) \le \tilde{C}$ mes Γ where $\tilde{C} > 0$ is independent of ϵ and Γ. It follows from here that any weak limit μ of μ^ϵ is absolutely continuous with respect to mes. Since F is ergodic with respect to mes and μ is also F-invariant then μ must coincide with mes.

Next, we shall specify the model of random perturbations we shall deal with in this chapter.

Assumption 1.1. (a) Transition probabilities $P^\epsilon(x,\cdot)$ of Markov chains X_n^ϵ have the form $P^\epsilon(x,\cdot) = Q_{Fx}^\epsilon(\cdot)$ where a family of measures $Q_y^\epsilon \in \mathcal{P}(M)$, $y \in M$ have densities q_y^ϵ with respect to the Riemannian volume m on M, i.e., $Q_y^\epsilon(\Gamma) = \int_\Gamma q_y^\epsilon(z)dm(z)$ for any Borel set $\Gamma \subset M$;

(b) There exist constants α, $C > 0$, $\alpha < 1$ and a family of non-negative functions $\{r_z(\xi), x \in M, \xi \in T_xM\}$, where T_xM denotes the tangent space at x, such that

$$q_x^\epsilon(y) \le C\epsilon^{-\nu} e^{-\frac{\alpha}{\epsilon}\mathrm{dist}(x,y)} \quad \text{for all } x,y \in M, \tag{1.6}$$

and

$$q_x^\epsilon(y) \le (1+\epsilon^\alpha)\epsilon^{-\nu} r_x(\tfrac{1}{\epsilon}\mathrm{Exp}_x^{-1} y) \tag{1.7}$$

provided $\mathrm{dist}(x,y) \le \epsilon^{1-\alpha}$, where $\mathrm{Exp}_x : T_xM \to M$ is the exponential map, $\epsilon > 0$ is small enough, and we suppose that the injectivity radius ρ_M of M is positive, i.e., for any $x \in M$ the maps Exp_x is a diffeomorphism of the ρ_M-ball in T_xM centered at $0 \in T_xM$ with the ρ_M-ball in M centered at x;

(c) The functions $r_x(\xi)$, $x \in M$, $x \in T_xM$ satisfy

(i) $\int_{T_xM} r_x(\xi)dm_x(\xi) = 1$,

where m_x is the volume induced by the Riemannian metric in T_xM;

(ii) $r_x(\zeta) \leq Ce^{-\alpha\|\xi\|}$,

where $\|\cdot\|$ denotes the norm in the corresponding tangent space induced by the Riemannian metric and $\alpha, C > 0$ are independent of x and ξ;

(iii) There exists $C > 0$ such that if $V_x^+ = \{\xi \in T_xM : r_x(\xi) > 0\}$ and $\partial V_x^+(\delta)$ denotes the δ-neighborhood in T_xM of the boundary ∂V_x^+ of V_x^+ then

$$\int_{\partial V_x^+(\delta)} r_x(\xi)dm_x(\xi) \leq C\delta, \tag{1.8}$$

and

$$r_x(\xi) \leq r_y(\eta) + C\rho + \chi_{\partial V_x^+(C\rho)}(\xi)r_x(\xi) \tag{1.9}$$

where $\rho = \rho((x,\xi),(y,\eta)) = \text{dist}(x,y) + \|\xi - \pi_{yx}\eta\|$, π_{yx} is the parallel displacement from T_yM to T_xM which is defined provided $\text{dist}(x,y)$ does not exceed the injectivity radius ρ_M, and, as usual, χ_Γ denotes the indicator function of a set Γ.

Remark 1.1. The condition (iii), though looking complicated, enables us to consider the functions $r_x(\xi)$ with compact support V_x^+ having discontinuity on ∂V_x^+ and satisfying a kind of the Lipschitz condition inside of V_x^+. It includes also some continuity of V_x^+ in x. The functions $r_x(\xi)$ having compact supports generate models of local random perturbations (see Example 1.2 below). Note, also, that both (1.6) and (c)(ii) can be relaxed to a decreasing with some polynomial speed.

Example 1.2. Sinai [Si1] suggested the following model of local random perturbations which corresponds to a particle jumping from x to a point distributed randomly in the ε-ball centered at Fx. In this model $r_x(\xi) = 0$ for $\xi \in T_xM$ with $\|\xi\| > \rho_x$ where $\rho_x > 0$ is a smooth function of $x \in M$. When $\|\xi\| < \rho_x$ then $r_x(\xi)$ is supposed to be nonnegative and smooth in both x and ξ in a small neighborhood of x and ξ while we stay away from the boundaries of ρ_x-balls in T_xM. These generate transition probabilities $P^\varepsilon(x,\cdot)$ as in Assumption 1.1 supported by $\overline{U_{\varepsilon\rho_x}(x)} = \{x: \mathrm{dist}(x,z) \le \varepsilon\rho_x\}$.

Example 1.3. One obtains another important model introduced by the author in [Ki1] and [Ki2] assuming that a particle jumps from x to Fx and then performs a diffusion for the time ε^2. This means that $X^\varepsilon_{n+1} = \xi_\varepsilon(FX^\varepsilon_n)$, where $\xi_t(x)$ is a solution of an Ito stochastic differential equation (see Friedman [Fri] vol. 1, Ikeda and Watanabe [IW], and Chapter III of this book)

$$d\xi_t(x) = \sigma(\xi_t(x))dw_t + b(\xi_t(x))dt$$

with the initial condition $\xi_0(x) = x$. Here $\sigma(y)$ and $b(y)$ are smooth $\nu\times\nu$-matrix and ν-vector functions, respectively, and w_t is a ν-dimensional standard Wiener process (Brownian motion). Stochastic differential equations can be considered on manifolds, as well, (see [IW]) by piecing together the solutions in different coordinate neighborhoods. Transition probabilities of the diffusion process ξ_t have densities $p(t,x,y)$ which are the fundamental solutions of the parabolic equation (called Kolmogorov's equation) having the form $\frac{\partial p(t,x,y)}{\partial t} =$ $Lp(t,x,y)$, where

$$L = \frac{1}{2}\langle a(x)\nabla_x, \nabla_x\rangle + \langle b(x), \nabla_x\rangle,$$

$a(x) = \sigma(x)\sigma^*(x)$, ∇_x is the gradient acting in x, and $\langle\cdot,\cdot\rangle$ denotes the inner product. The matrix field $a(x) = (a_{ij}(x))$ on the manifold M generates a Riemannian metric with the length element $\sum_{i,j} a^{ij}(x)dx_i dx_i$, $(a^{ij}) = (a_{ij})^{-1}$. Denote the corresponding norm in tangent spaces by $\|\cdot\|_a$. It is not difficult to show by the parametrix method (see Appendix to this chapter) that if $\mathrm{dist}(y,z) \le \epsilon^{1-\alpha}$ with $\alpha < \frac{1}{3}$ then

$$\left|1-p(\epsilon^2,y,z)\left[(2\pi\epsilon^2)^{-\frac{\upsilon}{2}}(\det a(y))^{-\frac{1}{2}}\exp\left[-\frac{1}{2\epsilon^2}\|\mathrm{Exp}_y^{-1}z\|_a^2\right]\right]^{-1}\right|$$

$$\le C\epsilon^{1-3d}, \tag{1.10}$$

for some $C > 0$ independent of y, z, and $\epsilon > 0$. Thus Assumption 1.1 will be satisfied if we take $r_x(\xi)$ $= (2\pi)^{-\upsilon/2}(\det a(x))^{-1/2}\exp\left[-\frac{1}{2\epsilon^2}\|\xi\|_a^2\right]$, $q_y^\epsilon(z) = p(\epsilon^2,y,z)$, and transition probabilities $P^\epsilon(x,\Gamma) = \int_\Gamma p(\epsilon^2,Fx,z)dm(z)$ of the Markov chains X_n^ϵ.

Example 1.4. The following example is the main model of random perturbations of invertible dynamical sytems F^t with continuous time, $-\infty < t < \infty$, i.e., flows. Suppose that a flow F^t is given by a vector field $B(x)$ on M so that

$$\frac{dF^t x}{dt} = B(F^t x), \quad F^0 x = x. \tag{1.11}$$

Let L be a second order elliptic differential operator on M with smooth coefficients. Such operators can always be represented in the form $L = \frac{1}{2}\Delta^{(a)} + b$ where b is a vector field and $\Delta^{(a)} = \mathrm{div}^{(a)}\mathrm{grad}^{(a)}$ is the Laplace-Beltrami operator corresponding to some metric form $\sum_{i,j} a^{ij}(x)dx_i dx_j$ which determines our Riemannian metric here. Consider the new operator $L^\epsilon = \epsilon^2 L + B$ which generates a Markov diffusion process X_t^ϵ with transition probabilities $P^\epsilon(t,x,\cdot)$ having densities $p^\epsilon(t,x,y)$ with respect to the Riemannian volume satisfying Kolmogorov's equation

$$\frac{\partial p^\epsilon(t,x,y)}{\partial t} = L^\epsilon p^\epsilon(t,x,y)$$

where L^ϵ acts in the variable x (see Friedman [Fri], vol. 1 and Ikeda and Watanabe [IW]). In Appendix to this chapter we shall show by the parametrix method that if $\mathrm{dist}(F^t x, y) \leq \epsilon^{1-\alpha}$ with $\alpha < \frac{1}{3}$, $0 < t \leq 1$, then for some $C > 0$,

$$|1-p^\epsilon(t,x,y)\left[\epsilon^{-\nu} r^t_{F^t x}\left[\frac{1}{\epsilon}\mathrm{Exp}^{-1}_{F^t x}\, y\right]\right]^{-1}| \leq C\epsilon^{1-3\alpha}, \tag{1.12}$$

where

$$r_y(t,\xi) = (2\pi)^{-\frac{\nu}{2}}(\det{}_{F^t y} A_t)^{-\frac{1}{2}} \exp(-\tfrac{1}{2}\langle A_t^{-1}\xi,\xi\rangle), \tag{1.13}$$

$$\xi \in T_{F^t y} M, \quad A_t = \int_0^t DF^u (DF^u)^* du,$$

the differential DF^u acts from $T_y M$ to $T_{F^u y} M$ and the adjoint operator $(DF^u)^*$ acts from $T_{F^u y} M$ to $T_y M$ so that A_t is the linear automorphism of $T_{F^t y}M$ and its determinant $\det_{F^t y}$ can be defined in a correct way. Taking $r_y(\xi) = r_y(1,\xi)$ and $q_y^\epsilon(z) = p^\epsilon(1,F^{-1}y,z)$ we obtain Markov chains X_n^ϵ with transition probabilities $P^\epsilon(x,\cdot) = P^\epsilon(1,x,\cdot)$ which are (diffusion type) random perturbations of the time-one map $F = F^1$ of the flow F^t satisfying Assumption 1.1. Since invariant measures of Markov chains X_n^ϵ are also invariant for diffusion processes X_t^ϵ, we may consider only the discrete-time case. In view of the relation (I.2.19) this remark concerns also the approximation of the entropy via random perturbations.

Next, we shall see that Assumption 1.1 enables us to restrict our attention to random paths of Markov chains X_n^ϵ which are δ-pseudo-orbits.

Lemma 1.1. *If* M *is not compact suppose that the volume of balls in* M *grows with radius* ρ *not faster than* Ce^ρ. *Then Assumption 1.1(a) together with (1.5) imply*

$$|P^\epsilon(n,x,\Gamma)-P_x^\epsilon\{\text{dist}(FX_i^\epsilon,X_{i+1}^\epsilon) < \delta \tag{1.14}$$

$$\text{for all } i = 0,\dots,n-1 \text{ and } X_n^\epsilon \in \Gamma\}| \le Cn\epsilon^{-2\nu}m(\Gamma)e^{-\frac{\alpha\delta}{2\epsilon}}$$

for some $C > 0$ *and any* $\delta > 0$, *an integer* $n > 0$, $\epsilon > 0$ *small enough, and a Borel set* $\Gamma \subset M$.

Proof. By (1.6) and the Markov property the left hand side of (1.14) does not exceed the expression

$$\sum_{i=0}^{n-1} P_x^\epsilon\{\text{dist}(FX_i^\epsilon, X_{i+1}^\epsilon) \geq \delta \text{ and } X_n^\epsilon \in \Gamma\} \tag{1.15}$$

$$\leq (n-1)\sup_y P_y^\epsilon\{\text{dist}(Fy, X_1^\epsilon) \geq \delta\}C\epsilon^{-\upsilon}m(\Gamma)$$

$$+ \sup_y P_y^\epsilon\{\text{dist}(Fz, X_1^\epsilon) \geq \delta \text{ and } X_1^\epsilon \in \Gamma\}$$

$$= (n-1)C\epsilon^{-\upsilon}m(\Gamma)\sup_y \int_{M\setminus U_\delta(Fy)} q_{Fy}^\epsilon(z)dm(z)$$

$$+ \sup_y \int_{M\setminus U_\delta(Fy)} q_{Fy}^\epsilon(z)dm(z)$$

which in view of (1.6) implies (1.14). □

Corollary 1.1. *Let* $\Lambda \subset M$ *be an attractor having a compact neighborhood. Then for any open set* $V \supset \Lambda$ *there exist numbers* $r, \beta, \epsilon_0 > 0$ *such that for all* $N = 1,2,\ldots$ *one has*

$$P_x^\epsilon\{\tau_{M\setminus V} < N\} < N^2 e^{-\frac{\beta}{\epsilon}} \tag{1.16}$$

provided $x \in U_r(\Lambda)$ *and* $0 < \epsilon < \epsilon_0$.

Proof is the same as Corollary I.5.1 by employing (1.14) and the fact that any δ-pseudo-orbit starting in $U_r(\Lambda)$ remains in V forever provided $\delta > 0$ and $r > 0$ are small enough. □

Corollary 1.1 together with (1.2) enables us to restrict our attention to the study of transition probabilities $P^\epsilon(n,x,\Gamma)$ while Markov chain X_k^ϵ remains for all $k = 0,1,\ldots,n$ in a neighborhood of an attractor, provided n is not too big, so that the right hand side of (1.16) is small. In fact, we shall use the time n not exceeding $\epsilon^{-\alpha}$, $\alpha > 0$ (in fact, $n \sim (\log \epsilon)^2$) since our estimates will require n applications of (1.7) yielding the multiplication by $(1+\epsilon^{\alpha})^n$ which must be bounded in order to arrive to our absolute continuity conclusions. These remarks justify the following.

Assumption 1.2. There exists an attractor $\Lambda \subset M$ with an open neighborhood $U_\Lambda \supset \Lambda$ satisfying (I.4.2) and having a compact closure $\bar{U}_\Lambda$ such that $\overline{FU_\Lambda} \subset U_\Lambda$, and

$$q_x^{\epsilon}(y) = 0 \quad \text{if} \quad x \in \overline{FU_\Lambda} \quad \text{and} \quad y \notin U_\Lambda. \tag{1.17}$$

Remark that if $U \supset \Lambda$ satisfies $F^n U \subset V$ for some $n > 0$ and an open set V such that $\bar{V} \subset U$ then $U_\Lambda = \bigcap_{i=0}^{n-1} F^i U$ satisfies both (I.4.2) and $\overline{FU_\Lambda} \subset U_\Lambda$. On this stage Assumption 1.2 is only a compactness condition. It enables us to restrict our study to the neighborhood U_Λ since Markov chains X_n^{ϵ} never leave $\bar{U}_\Lambda$ once starting there. Since $\bar{U}_\Lambda$ is compact then under Assumption 1.2 the Markov chains X_n^{ϵ} possess invariant measures supported by $\bar{U}_\Lambda$. Remark that Assumption 1.1 does not imply any uniqueness of these measures since, for instance, the conditions of Proposition I.1.8 may be still not satisfied.

Our method will rely heavily on the following condition which substitutes the study of Markov chains X_n^{ϵ} along δ-pseudo-orbits (see (I.4.1)) by the study of X_n^{ϵ} along true orbits of F.

Definition 1.1. A transformation F is said to have the shadowing property with parameters $\beta \geq 0$ and $C > 0$ on a set $V \subset M$ if there is $\delta_0 > 0$ such that for all $n = 1,2,\ldots$ and each δ-pseudo-orbit $\omega = (z_0,\ldots,z_n)$ with $\delta \leq \delta_0$ and $z_i \in V$, $i = 0,1,\ldots,n$ one can find a point $y_\omega \in V$ satisfying

$$\mathrm{dist}(F^i y_\omega, z_i) \le Cn^\beta \delta \quad \text{for all} \quad i = 0,1,\dots,n. \tag{1.18}$$

There exist other versions of the shadowing property called also the pseudo-orbit tracing property refering to the fact that pseudo-orbits are shadowed or traced by true orbits. Usually, the specific form of the right hand side of (1.18) does not play essential part in versions of the shadowing property needed for the deterministic theory of stability of dynamical systems (see, for instance, Hirsch, Pugh and Shub [HPS] and Shub [Sh]), though it will be important for our study. We shall see how to establish the shadowing property for transformations F satisfying certain hyperbolicity or expanding conditions. This property written in the form (1.18) holds both for hyperbolic diffeomorphisms and hyperbolic flows.

Example 1.5. Take F to be an automorphism of the two-dimensional torus $\mathcal{T}^2$ generated by the matrix $\begin{pmatrix} 2 & 1 \\ 1 & 1 \end{pmatrix}$. This matrix has two positive eigenvalues $\lambda_1 = \frac{3+\sqrt{5}}{2} > 1$ and $\lambda_2 = \frac{3-\sqrt{5}}{2} < 1$ with the eigenvectors $\tilde{e}_1 = \left(\frac{1+\sqrt{5}}{2}, 1\right)$ and $e_2 = \left(\frac{1-\sqrt{5}}{2}, 1\right)$, respectively. An application of F stretches all vectors parallel to e_1 and contracts all vectors parallel to e_2. Now let $x_0, x_1, \dots, x_n$ be a δ-pseudo-orbit of F on $\mathcal{T}^2$ with $\delta > 0$ small enough. Consider the squares A_i centered at x_i, $i = 0,1,\dots,n$ with sides of the length $\delta(\sqrt{5}+1)$ parallel to e_1 and e_2. It is easy to see that the rectangle FA_i stretches across A_{i+1} intersecting only the part of its boundary which is parallel to e_2 as shown on the following picture.

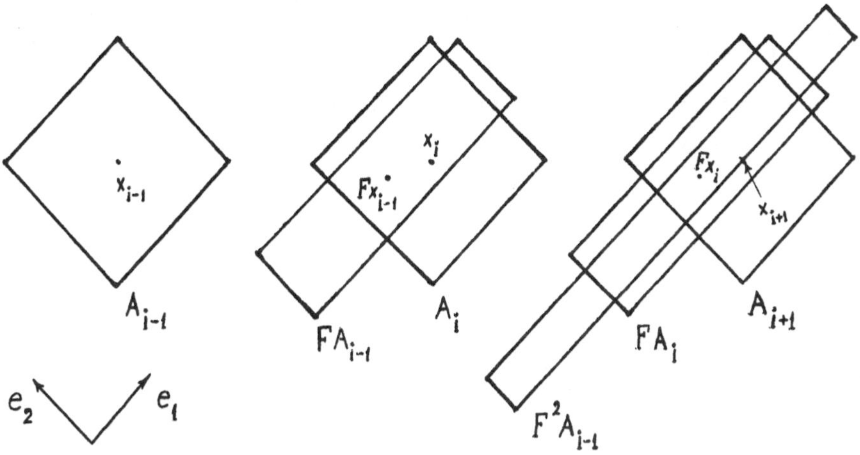

Figure 1.1

It follows from this that the intersection $\Gamma = \bigcap_{i=0}^{n} F^{n-i} A_i$ is not empty and any point y from $F^{-n}\Gamma$ satisfies $F^i y \in A_i$, $i = 0,1,\dots,n$, and so

$$\mathrm{dist}(F^i y, x_i) \leq \delta \frac{(\sqrt{5}+1)}{2} .$$

Thus for this case the shadowing property holds true with $V = \mathcal{T}^2$, $\beta = 0$, and $C = \frac{\sqrt{5}+1}{\sqrt{2}}$.

In Section 2.3 we shall establish the shadowing property for more general hyperbolic and expanding transformations.

2.2. Markov chains in tangent bundles.

In this section we shall construct certain Markov chains in the tangent bundle TM which will play an important part in our study. These Markov chains can be considered as a linearization of random perturbations X_n^ϵ along the orbits of the transformation F.

For any $\xi \in T_zM$, $z \in M$ and a Borel set $\Psi \subset TM$ define

$$R_z^\epsilon(\xi,\Psi) = \int_{\Psi\cap T_F M} r_z^\epsilon(\xi,\eta)dm_{Fz}(\eta) \tag{2.1}$$

where

$$r_z^\epsilon(\xi,\eta) = \epsilon^{-\nu} r_{Fz}\left[\frac{\eta-DF\xi}{\epsilon}\right], \tag{2.2}$$

m_y is the Riemannian volume in T_yM and the functions $\{r_y(\zeta)\}$ where introduced in Assumption 1.1, and DF is the differential of F.

Lemma 2.1. *Let* $\{\theta_x(k) \in T_{F^kx}M, k = 1,2,\ldots\}$ *be mutually independent random vectors with the distributions*

$$P\{\theta_x(k) \in \Psi\} = \int_\Psi r_{F^kx}(\eta)dm_{F^kx}(\eta) \tag{2.3}$$

for any Borel $\Psi \subset T_{F^kx}M$. *Then*

$$\Xi_x^\epsilon(\zeta,n) = \epsilon \sum_{k=1}^{n} DF^{n-k}\theta_x(k) + DF^n\zeta \tag{2.4}$$

$n = 0,1,\ldots$ *is a nonhomogeneous Markov chain starting at* ζ *and having transition probabilities* $R_{F^nx}^\epsilon(\xi,\cdot)$ *provided* $\Xi_x^\epsilon(\zeta,n-1) = \xi$.

Proof. Since $\theta_x(n)$ and $\Xi_x^\epsilon(\zeta,n-1)$ are independent it follows that

$$P\{\Xi_x^\epsilon(\zeta,n) \in \Psi | \Xi_x^\epsilon(\zeta,n-1)\}$$

$$= P\{\epsilon\theta_x(n) + DF\Xi_x^\epsilon(\zeta,n-1) \in \Psi | \Xi_x^\epsilon(\zeta,n-1)\}$$

$$= R^\epsilon_{F^n x}(\Xi_x^\epsilon(\zeta,n-1),\Psi)$$

where, as usual, $P\{\cdot|\cdot\}$ denotes the conditional probability. This proves our assertion. □

For $\xi \in T_xM$ and a Borel set $\Psi \subset T_{F^n x}M$ denote by $R_x^\epsilon(n,\xi,\Psi)$ the probability of the event $\{\Xi_x^\epsilon(\xi,n) \in \Psi\}$. By the Chapman-Kolmogorov formula (see (I.1.19)) and Lemma 2.1 one has

$$R_x^\epsilon(n,\xi,\Psi) = \int_{T_{Fx}M} \cdots \int_{T_{F^{n-1}x}M} \int_\Psi r_x^\epsilon(\xi,\eta_1) r_{Fx}^\epsilon(\eta_1,\eta_2) \tag{2.5}$$

$$\cdots r^\epsilon_{F^n x}(\eta_{n-1},\eta_n) dm_{Fx}(\eta_1)\cdots dm_{F^n x}(\eta_n).$$

In this section we do not use, in fact, the tangent bundle structure and all results will remain true if we replace $T_{F^k x}M$, $k = 0,1,\ldots$ and $DF: T_{F^k x}M \to T_{F^{k+1}x}M$ by a sequence of Euclidean spaces L_k, $k = 0,1,\ldots$ (which, actually, can be identified) and a corresponding sequence of invertible linear maps $A_k: L_k \to L_{k+1}$. Still, we shall stick to the tangent bundle notations in view of our applications where we shall have certain DF-invariant splittings of the tangent bundle TM over, so called, hyperbolic sets into stable (exponentially contracting) and unstable (exponentially expanding) subbundles. Keeping this in mind we shall obtain first the following result which will be important in Section 2.4 for the proof of absolute continuity of limiting measures of random perturbations in the unstable direction.

Proposition 2.1. *Let* $F:M \to M$ *be a smooth endomorphism such that for some* $x \in M$ *there is a splitting* $T_{F^i x} M = \tilde{H}_{F^i x} \oplus H^u_{F^i u}$ (u *stands for unstable*), $i = 0,1,2,...$ *satisfying* $DF\tilde{H}_{F^i x} = \tilde{H}_{F^{i+1}x}$ *and* $DFH^u_{F^i x} = H^u_{F^{i+1}x}$, *the angle between the subspaces* $\tilde{H}_{F^i x}$ *and* $H^u_{F^i x}$ *is bounded away from zero uniformly in* i, *and there exists* $\gamma > 0$ *so that for each* $\zeta \in H^u_{F^k x}$ *and* $k \geq \ell \geq 0$,

$$\|DF^{-\ell}\zeta\| \leq \gamma^{-1} e^{-\gamma\ell} \|\zeta\|. \tag{2.6}$$

Then there exist $\sigma, C > 0$ *such that for any* $\xi \in T_x M$, $n \geq 1$, *Borel sets* $\tilde{\Psi} \subset \tilde{H}_{F^n x}$ *and* $\Psi^u \subset H^u_{F^n x}(1) = \{\zeta \in H^u_{F^n x} : \|\zeta\| \leq 1\}$ *one has*

$$R^\epsilon_x(n,\xi,\tilde{\Psi}+\Psi^u) \leq C\epsilon^{-\nu^u} m^u_{F^n x}(\Psi^u) \mathcal{J}_n^{-1}(x) \tag{2.7}$$

where $\mathcal{J}_n(x)$ *is the absolute value of the Jacobian of the linear map* $DF^n : H^u_x \to H^u_{F^n x}$, $m^u_{F^i x}$ *is the volume on* $H^u_{F^i x}$ *induced by* $m_{F^i x}$, ν^u *is the dimension of* H^u_x, *and* $\tilde{\Psi}+\Psi^u = \{\zeta+\eta : \zeta \in \tilde{H}_{F^n x}, \eta \in H^u_{F^n x}\}$. *If in addition,* $n \geq (\log \epsilon)^2$ *and* ϵ *is small enough then*

$$R_x^\epsilon(n,\xi,\tilde{\Psi}+\Psi^u) \le C\epsilon^{-\nu^u} m^u_{F^n x}(\Psi^u)\mathcal{J}_n^{-1}(x)e^{-\sigma\|\xi^u\|\epsilon^{-1}} \tag{2.8}$$

where $\xi^u = \pi\xi$ *is the projection of* ξ *on* H_x^u *parallel to* $\tilde{H}_x$.

Proof. Put

$$\varphi_x^u(n) = \sum_{k=1}^{n-1} DF^{-k}\theta_x(k+1). \tag{2.9}$$

Then

$$R_x^\epsilon(n,\xi,\tilde{\Psi}+\Psi^u) \le P\{\epsilon DF^{n-1}(\varphi_x^u(n)+\theta_x^u(1))+DF^n\xi^u \in \Psi^u\} \tag{2.10}$$

$$= P\{\varphi_x^u(n)+\theta_x^u(1)+\epsilon^{-1}DF\xi^u \in \epsilon^{-1}DF^{-(n-1)}\Psi^u\}$$

$$= \int_{H^u_{Fx}} P\{\varphi_x^u(n) \in d\eta\} \int_{\epsilon^{-1}DF^{-(n-1)}\Psi^u} r^u_{Fx}(\zeta-\eta-\epsilon^{-1}DF\xi^u)dm^u_{Fx}(\zeta)$$

since $\varphi_x^u(n)$ and $\theta_x^u(1)$ are independent, where for $y = F^i x$,

$$r_y^u(\eta) = \int_{\tilde{H}_y} r_y(\eta+\zeta)\tilde{m}_y(d\zeta) \tag{2.11}$$

is the density with respect to m_y^u of the distribution of $\theta_y(i)$, and $\tilde{m}_y$ is the volume on $\tilde{H}_y$ induced by m_y. In view of the property (ii) of the functions $\{r_y(\eta)\}$ in Assumption 1.1 it follows from (2.10) and (2.11) that for some $C > 0$, $R_x^\epsilon(n,\zeta,\tilde{\Psi}+\Psi^u) \le Cm^u_{Fx}(\epsilon^{-1}DF^{-(n-1)}\Psi^u)$ implying (2.7) by the definition of the Jacobian $\mathcal{J}_n(x)$.

Next, it is easy to see that for some constants $\tilde{\sigma}$, $\tilde{C} > 0$,

$$r_y^u(\eta) \le \tilde{C}\exp(-\tilde{\sigma}\|\eta\|),\ y = F^i x,\ i = 0,1,\dots, \tag{2.12}$$

and

$$E \exp(\tilde{\sigma}\|\varphi_x^u(n)\|) \le \tilde{C} \tag{2.13}$$

where E denotes the expectation. Indeed, the inequality (2.12) follows from (2.11), the property (ii) in Assumption 1.1, and the uniform transversality of $\tilde{H}_{F^i x}$ and $H^u_{F^i x}$.

Next, by (2.6),

$$\|\varphi_x^u(n)\| \le \gamma^{-1} \sum_{k=1}^{n-1} e^{-\gamma k}\|\theta_x^u(k+1)\|. \tag{2.14}$$

Since $\{\theta_x^u(k), k = 1,2,...\}$ are mutually independent then by (2.14), the property (ii) in Assumption 1.1, and the uniform transversality of $\tilde{H}_{F^k x}$ and $H^u_{F^k x}$ it follows that for $\tilde{\sigma} > 0$ small enough

$$E \exp(\tilde{\sigma}\varphi_x^u(n)) \le \prod_{k=1}^{n} E \exp\{\tilde{\sigma}\gamma^{-1}e^{-\gamma k}\|\theta_x^u(k+1)\|\} \tag{2.15}$$

$$\le \prod_{k=1}^{n} E \exp\{\tilde{\sigma}\gamma^{-1}e^{-\gamma k}\tilde{C}\|\theta_x(k+1)\|\}$$

$$\le \prod_{k=1}^{n} (\exp\{\tilde{\sigma}\gamma^{-1}\tilde{C}e^{-\frac{1}{2}\gamma k}\}$$

$$+ \int_{\{\zeta:\|\zeta\|>e^{\frac{1}{2}\gamma k}\}} \exp\{(\tilde{\sigma}\gamma^{-1}\tilde{C}e^{-\gamma k}-\alpha)\|\zeta\|\}dm_{F^{k+1}x}(d\zeta)$$

implying (2.13), where $\tilde{C}^{-1} = \min_i \sin$ of the angle between $\tilde{H}_{F^i x}$ and $H^u_{F^i x}$. Now by the Chebyshev inequality we derive from (2.13) that

$$P\{\|\varphi_x^u(n)\| \ge \frac{1}{3\epsilon}\|DF\xi^u\|\} \le \tilde{C}\exp(-\frac{\tilde{\sigma}}{3\epsilon}\|DF\xi^u\|). \tag{2.16}$$

If $n \geq (\log \epsilon)^2$ and $\epsilon > 0$ is small enough then

$$\sup\{\|\zeta\| : \zeta \in \epsilon^{-1} DF^{-(n-1)} \Psi^u\} \leq 1. \tag{2.17}$$

Hence if $\zeta \in \epsilon^{-1} DF^{-(n-1)} \Psi^u$ and $\|\eta\| < \frac{1}{3}\epsilon \|DF\xi^u\|$ then by (2.12),

$$r^u_{Fx}(\zeta - \eta - \epsilon^{-1} DF\xi^u) \leq \tilde{C} \exp(-\frac{\tilde{\sigma}}{3\epsilon} \|DF\xi^u\|). \tag{2.18}$$

Representing the integral over H^u_{Fx} in (2.10) as the sum of the integral over the set $\{\eta \in H^u_{Fx} : \|\eta\| \geq \frac{1}{3}\epsilon \|DF\xi^u\|\}$ and the integral over its complement in H^u_{Fx} we derive from (2.10), (2.12), (2.16), and (2.18) for ϵ small enough that

$$R^\epsilon_x(n, \xi, \tilde{\Psi} + \Psi^u) \tag{2.19}$$

$$\leq \tilde{C}(\tilde{C}+1) \exp(-\frac{\tilde{\sigma}}{3\epsilon} \|DF\xi^u\|) m^u_{Fx}(\epsilon^{-1} DF^{-(n-1)} \Psi^u)$$

implying (2.8). □

Under additional conditions we are able to obtain also an estimate from below of the probability $R^\epsilon_x(n, \xi, \Psi)$. This and the whole remaining part of this section will be needed only in Section 2.7 concerning the topological pressure.

Proposition 2.2. *In the circumstances of Proposition 2.1 suppose, in addition, that* $r_{Fx}(\eta) > 0$ *if* $\eta \in T_{Fx}M$ *and for any* $\zeta \in \tilde{H}_{F^k x} = H^s_{F^k x}$ (s stands for stable), *and* $k, \ell \geq 0$,

$$\|DF^\ell \zeta\| \leq \gamma^{-1} e^{-\gamma \ell} \|\zeta\|. \tag{2.20}$$

Assume also that $H^s_{F^n x} \supset \Psi = \Psi^s \supset H^s_{F^n x}(\delta)$ $= \{\zeta \in H^s_{F^n x} : \|\zeta\| \le \delta\}$, $\xi = \xi^s + \xi^u$ *where* $\xi^s \in H^s_x$, $\|\xi^s\| \le 1$ *and* $\xi^u \in H^u_x$, $\|\xi^u\| \le \epsilon$. *Then*

$$R^\epsilon_x(n,\xi,\Psi^s+\Psi^u) \ge C\epsilon^{-v^u} m^u_{F^n x}(\Psi^u)\mathcal{J}^{-1}_n(x) \tag{2.21}$$

$$-C^{-1}\exp(-C^{-1}\delta\epsilon^{-1})$$

for some $C > 0$ *independent of* ϵ, δ, n, *provided* $n \ge (\log \epsilon)^2$ *and* ϵ *is small enough.*

Proof. Put

$$\varphi^s_x(n) = \sum_{k=1}^{n} DF^{n-k}\theta^s_x(k) \tag{2.22}$$

where we use the unique decomposition $\theta_x(k) = \theta^s_x(k)+\theta^u_x(k)$ with $\theta^s_x(k) \in H^s_{F^k x}$ and $\theta^u_x(k) \in H^u_{F^k x}$. Then

$$R^\epsilon_x(n,\xi,\Psi+\Psi^u) \tag{2.23}$$

$$\ge P\{\epsilon DF^{n-1}(\varphi^u_x(n)+\theta^u_x(1)+DF^n\xi^u) \in \Psi^u\}$$

$$-P\{\|\epsilon\varphi^s_x(n)+DF^n\xi\| \ge \delta\}.$$

Since $\theta_x(k)$, $k = 1,2,\ldots$ are independent then by (2.20), the property (ii) in Assumption 1.1, and the uniform transversality of $H^s_{F^k x}$ and $H^u_{F^k x}$ in the same way as in (2.13) it follows that

$$E\exp(\tilde{\sigma}\|\varphi^s_x(n)\|) \le \tilde{C} \tag{2.24}$$

for some $\tilde{\sigma},\tilde{C} > 0$ independent of n. Since $\|\xi^s\| \leq 1$ and $n \geq (\log \epsilon)^2$ then by (2.20), $\epsilon^{-1}\|DF^n\xi^s\| \leq 1$ provided ϵ is small enough. Thus by the Chebyshev inequality

$$P\{\|\epsilon\varphi_x^s(n)+DF^n\xi^s\| \geq \delta\} \leq P\{\|\varphi_x^s(n)\| \geq \delta\epsilon^{-1}-1\} \tag{2.25}$$

$$\leq \tilde{C}\exp(\tilde{\sigma}(1-\delta\epsilon^{-1})).$$

In the same way as in (2.10),

$$P\{\epsilon DF^{n-1}(\varphi_x^u(n)+\theta_x^u(1))+DF^n\xi^u \in \Psi^u\} \tag{2.26}$$

$$= \int_{H_{Fx}^u} P\{\varphi_x^u(n) \in d\eta\} \int_{\epsilon^{-1}DF^{-(n-1)}\Psi^u} r_{Fx}^u(\zeta-\eta-\epsilon^{-1}DF\xi^u)dm_{Fx}^u(\xi)$$

$$\geq \frac{1}{2}\epsilon^{-\nu^u} m_{Fx}^u(DF^{-(n-1)}\Psi^u) \inf_{\zeta\in T_{Fx}(1+\rho+\|DF\|_x)} r_{Fx}^u(\zeta)$$

where $T_y(q) = \{\eta \in T_yM: \|\eta\| \leq q\}$ and $\rho = \inf\{q: P\{\varphi_x^u(n) \in T_{Fx}(q)\} \geq \frac{1}{2}$ for all $n \geq 0\}$. It follows from (2.13) that $\rho \leq \tilde{\sigma}^{-1}\log 2\tilde{C}$ and since we suppose here r_{Fx} to be positive and continuous we conclude that the infinum in the right hand side of (2.26) is positive. This together with (2.25) gives (2.21). □

We have the following linear version of the shadowing property.

Proposition 2.3. *Let the splitting* $T_{F^kx}M = H_{F^kx}^s \oplus H_{F^kx}^u$, $k = 0,1,2,\ldots$ *satisfies assumptions of Propositions 2.1 and 2.2. Suppose that*

$$\tilde{D} = \sup_{k\geq 0}\|DF\|_{F^kx} < \infty \tag{2.27}$$

where $\|DF\|_y = \sup\{\|DF\eta\| : \eta \in H_y, \|\eta\| = 1\}$, $y = F^k x$, $k \geq 0$. *Then there exists a constant* $K > 0$ *such that for any collection of vectors* $\xi_i \in T_{F^i x} M$, $i = 0, \ldots, n$ *satisfying*

$$\|DF\xi_i - \xi_{i+1}\| \leq \delta, \ i = 0, \ldots, n-1 \tag{2.28}$$

one can find a vector $\eta \in T_x M$ *so that*

$$\|\xi_i - DF^i \eta\| \leq K\delta \quad \textit{for all} \quad i = 0, \ldots, n. \tag{2.29}$$

Proof. Pick up an integer n_0 so that $\gamma^{-1} e^{-\gamma n_0} < 1$ with γ from (2.6) and (2.20). Then $\zeta_0 = \xi_0$, $\zeta_1 = \xi_{n_0}, \ldots, \zeta_k = \xi_{kn_0}$, k = integral part of $(\frac{n}{n_0})$ is a $\tilde{C}\delta$-pseudo-orbit with $\tilde{C} = (\tilde{D}+1)^{n_0}$, i.e., $\|DF\zeta_j - \zeta_{j+1}\| \leq \tilde{C}\delta$, $j = 0, \ldots, k-1$. Then in the same way as in Example 1.5 (and, in fact, using the same picture) we conclude that if $\tilde{K} \geq 2\delta((1-r_0)\sin\varphi_{min})^{-1}$ where φ_{min} is the minimal angle between spaces H_y^s and H_y^u, $y = F^i x$, $i = 0, 1, \ldots, n$ then the intersection

$$Q = \bigcap_{j=0}^{k} DF^{jn_0}(\zeta_j + H^s_{F^{(k-j)n_0}x}(\tilde{K}\delta) + H^u_{F^{(k-j)n_0}x}(\tilde{K}\delta))$$

is not empty, where $H_y^s(q)$ and $H_y^u(q)$, $y = F^i x$ are q-balls in the corresponding subspaces centered at the origin. Then any point $\eta \in DF^{-kn_0} Q$ satisfies $\|DF^{jn_0}\eta - \zeta_j\| \leq \tilde{K}\delta$, $j = 0, 1, \ldots, k$ for some constant $\tilde{K} > 0$ depending only on φ_{min}, and so (2.29) will be satisfied with $K = \tilde{C}\tilde{K}$. □

Proposition 2.4. *Under the conditions of Proposition 2.3 there exists a constant* $C > 0$ *such that for any* $\delta > 0$, $\xi \in T_x(\delta) = \{\zeta \in T_xM: \|\zeta\| \le \delta\}$, $\Psi \subset T_{F^n x}(\delta)$, *and* $n \ge 0$,

$$|R_x^\epsilon(n,\xi,\Psi) - P\{\|\Xi_x^\epsilon(\xi,k)\| \le C\delta \text{ for all } k = 0,1,\dots,n \text{ and } \Xi_x^\epsilon(\xi,n) \in \Psi\}| \le Cn\epsilon^{-2\upsilon} m_{F^n x}(\Psi)\exp(-\frac{\alpha\delta}{2\epsilon}) \tag{2.30}$$

where α *is the same as in (ii) of* Assumption 1.1.

Proof. In the same way as in Lemma 1.1 we conclude that if one restricts the integration in (2.5) to δ-pseudo-orbits of DF then this may lead to a mistake of order of the right hand side of (2.30), i.e.,

$$|R_x^\epsilon(n,\xi,\Psi) - P\{\|DF\Xi_x^\epsilon(\xi,k) - \Xi_x^\epsilon(\xi,k+1)\| < \delta \text{ for all } k = 0,\dots,n-1 \text{ and } \Xi_x^\epsilon(\xi,n) \in \Psi\}| \le \tilde{C}n\epsilon^{-2\upsilon} m_{F^n x}(\Psi)\exp(-\frac{\alpha\delta}{2\epsilon}). \tag{2.31}$$

The δ-pseudo-orbits we have in (2.31) start at ξ with $\|\xi\| \le \delta$ and end in $\Psi \subset T_{F^n x}(\delta)$. According to Proposition 2.3 for each such δ-pseudo-orbit $\omega = (\xi_0,\xi_1,\dots,\xi_n)$ one can find a vector $\eta = \eta_\omega \in T_xM$ satisfying (2.29). In particular, $\|\eta\| \le (K+1)\delta$ and $\|DF^n\eta\| \le (K+1)\delta$. Representing $\eta = \eta^s + \eta^u$ and using (2.20) and (2.6) we derive from here that $\|DF^\ell\eta\| \le \tilde{\tilde{C}}(K+1)\delta$ for all $\ell = 0,1,\dots,n$ where a constant $\tilde{\tilde{C}} > 0$ does not depend on ω and η. This together with (2.29) yield $\|\xi_i\| \le (\tilde{\tilde{C}}(K+1)+K)\delta$. That is any δ-pseudo- orbit

$\omega = (\xi_0,\dots,\xi_n)$ of DF starting at δ-ball around zero in T_xM and ending in δ-ball around zero in $T_{F^nx}M$ has all points ξ_i belonging to $(\tilde{C}(K+1)+K)\delta$-balls around zero in $T_{F^ix}M$. Now taking $C = \max(\tilde{C},\tilde{C}(K+1)+K)$ we obtain (2.30).

□

Propositions 2.3 and 2.4 can be generalized in the following way.

Proposition 2.5. *Let* $T_{F^kx}M = H^s_{F^kx} \oplus H^0_{F^kx} \oplus H^u_{F^kx}$, $k = 0,1,\dots$ *where* $H^s_{F^kx}$ *and* $H^u_{F^kx}$ *satisfy the conditions of Propositions 2.1 and 2.2, all angles between the subspaces in this splitting are bounded away from zero uniformly in* k, $DFH^0_{F^kx} = H^0_{F^{k+1}x}$, *and the norm of the restriction* $DF^\ell|_{H^0_{F^kx}}$ *is bounded by a constant* $C > 0$ *independent of* $k,\ell \geq 0$. *Then there exists a constant* $K > 0$ *such that for any collection of vectors* $\xi_i \in T_{F^ix}M$, $i = 0,\dots,n$ *satisfying (2.28) one can find a vector* $\eta \in T_xM$ *so that*

$$\|\xi_i - DF^i\eta\| \leq Kn\delta \quad \text{for all} \quad i = 0,\dots,n. \tag{2.32}$$

Proof. Each vector $\zeta \in T_yM$, $y = F^kx$ has a unique representation $\zeta = \zeta^s+\zeta^0+\zeta^u$ with $\zeta^s \in H^s_y$, $\zeta^0 \in H^0_y$, and $\zeta^u \in H^u_y$. Denote $L_k = H^s_{F^kx} \oplus H^u_{F^ky}$. Then the sequence $\zeta_i = \xi^s_i+\xi^u_i$ will be a $\tilde{K}\delta$-pseudo-orbit in L_k, i.e., $\|DF\zeta_i - \zeta_{i+1}\| \leq \tilde{C}\delta$ for some $\tilde{C} > 0$ depending only on angles between subspaces. Then we can apply Proposition 2.3 finding some $\tilde{\eta} \in L_0$ satisfying

$$\|\zeta_i - DF^i\tilde{\eta}\| \le \tilde{K}\delta, \quad i = 0,\dots,n. \tag{2.33}$$

Put $\eta = \xi_0^0 + \tilde{\eta}$. Then

$$\|\xi_i^0 - DF^i\eta^0\| \le \sum_{j=1}^{i} \|DF^{i-j}\xi_j^0 - DF^{(i-j)+1}\xi_{j-1}^0\| \tag{2.34}$$

$$\le C \sum_{j=1}^{i} \|\xi_j^0 - DF\xi_{j-1}^0\| \le i\tilde{C}C\delta$$

where C is the bound for the norms $\|DF|_{H^0_{F^k x}}\|$, $k = 0,1,\dots$ and $\tilde{C} > 0$ depends only on angles between subspaces. Now (2.33) and (2.34) yield (2.32). □

Proposition 2.6. *In conditions of Proposition* 2.5 *there exists* $C > 0$ *such that for any* $\delta > 0$, $\xi \in T_x(\delta)$ *and* $\Psi \subset T_{F^n x}(\delta)$,

$$|R_x^\epsilon(n,\xi,\Psi) - P\{\|\Xi_x^\epsilon(\xi,k)\| \le Cn\delta \text{ for all } \tag{2.35}$$

$$k = 0,1,\dots,n \text{ and } \Xi_x^\epsilon(\xi,n) \in \Psi\}|$$

$$\le Cn\epsilon^{-2\upsilon} m_{F^n x}(\Psi)\exp(-\frac{\alpha\delta}{2\epsilon}).$$

Proof. We proceed in the same way as in Proposition 2.4 using Proposition 2.5 in place of Proposition 2.3. Restricting the integration in (2.5) to δ-pseudo-orbits of DF we make a mistake of order of the right hand side of (2.35). These δ-pseudo-orbits start at ξ with $\|\xi\| \le \delta$ and end in $\Psi \subset T_{F^n x}(\delta)$. All of them are shadowed with the precision $Kn\delta$ by orbits of vectors $\eta \in T_x M$ such that

$\|\eta\| \leq (K+1)\delta$ and $\|DF^n\eta\| \leq (K+1)\delta$. Representing $\eta = \eta^s+\eta^0+\eta^u$, using (2.20), (2.6), and $\|DF^i\eta^0\| \leq \tilde{C}\|\eta^0\|$, $i = 0,1,\ldots,n$ we derive $\|DF^i\eta\| \leq \tilde{\tilde{C}}\delta$ leading to (2.35). □

The lower bound (2.21) was proved in the case of hyperbolic splittings of diffeomorphisms and we are not able to extend it in the full generality to hyperbolic splittings of time-one maps of flows where one has an additional one-dimensional direction $H^0_{F^k x}$, $k = 0,1,\ldots$. Still, for the continuous time diffusion type random perturbations one obtains Markov processes in the tangent bundle with known transition probabilities given by (1.13) which was explained in Example 1.4 and will be proved in the Appendix to this chapter. For these transition probabilities one can obtain more precise estimates (cf. Kifer [Ki6], Section 4).

Define

$$r_z^\epsilon(t,\xi,\eta) = (2\pi\epsilon^2)^{-n/2}\left(\det_{F^t z} \int_0^t DF^\tau (DF^\tau)^* d\tau\right)^{-\frac{1}{2}} \tag{2.36}$$

$$\times \exp\left\{-\frac{1}{2\epsilon^2}\left\langle\left(\int_0^t DF^\tau (DF^\tau)^* d\tau\right)^{-1}(\eta - DF^t\xi), (\eta - DF^t\xi)\right\rangle_{F^t z}\right\}$$

where F^t is a C^1 flow, $\xi \in T_z M$, $\eta \in T_{F^t z} M$, $\langle\ ,\ \rangle_y$ is the inner product in $T_y M$ with respect to a fixed Riemannian metric, and the operator $\int_0^t DF^\tau (DF^\tau)^* d\tau$ transforms the space $T_{F^t z} M$ onto itself, and so the determinant $\det_{F^t z}$ of this operator is defined in a correct way using fixed Euclidean structure $\langle\ ,\ \rangle_{F^t z}$. Remark that in the case of diffusion type random perturbations one has to take a special metric connected with a differential operator as explained in Example 1.4. It is easy to see that

$$\int_{T_{F^t z}M} r_z^\epsilon(t,\xi,\eta)dm_{F^t z}(\eta) = 1, \ \xi \in T_z M, \ \eta \in T_{F^t z}M \tag{2.37}$$

and

$$\int_{T_{F^t z}M} r_z^\epsilon(t,\xi,\eta) r_{F^t z}^\epsilon(\tau,\eta,\zeta)dm_{F^t z}(\eta) = r_z^\epsilon(t+\tau,\xi,\zeta) \tag{2.38}$$

where $\xi \in T_z M$, $\zeta \in T_{F^{t+\tau}z}M$, and m_y is the Euclidean volume generated by the inner product $\langle \ , \ \rangle_y$.

According to (2.2) put

$$r_x^\epsilon(\xi,\eta) = r_x^\epsilon(1,\xi,\eta). \tag{2.39}$$

Then $r_x^\epsilon(k,\xi,\eta)$ is a k-step transition density of the Markov chain $\Xi_x^\epsilon(\xi,n)$ defined by (2.4), $F = F^1$, and so

$$R_x^\epsilon(n,\xi,\Psi) = \int_\Psi r_x^\epsilon(n,\xi,\eta)dm_{F^n x}(\eta). \tag{2.40}$$

Remark that

$$(2\pi\epsilon^2)^{-n/2}\left(\det_{F^t z} \int_0^t DF^\tau(DF^\tau)^* d\tau\right)^{-\frac{1}{2}} \tag{2.41}$$

$$\geq r_z^\epsilon(t,\xi,\eta) \geq (2\pi\epsilon^2)^{-n/2}\left(\det_{F^t z} \int_0^t DF^\tau(DF^\tau)^* d\tau\right)^{-\frac{1}{2}}$$

$$\times \exp\left\{-\epsilon^{-2}\left(\left\langle\left(\int_0^t DF^\tau(DF^\tau)^* d\tau\right)^{-1}\eta,\eta\right\rangle\right.\right.$$

$$\left.\left. + \left\langle\left(\int_0^t DF^{\tau-t}(DF^{\tau-t})^* d\tau\right)^{-1}\xi,\xi\right\rangle\right)\right\}$$

which will be the basis for our estimates.

Proposition 2.7. *Let the splitting* $T_{F^kx}M = H^s_{F^kx} \oplus H^0_{F^kx} \oplus H^u_{F^kx}$, $k = 0,1,\ldots$ *satisfies the conditions of Proposition 2.5 and the norm of the restriction* $DF^\ell\big|_{H^0_{F^kx}}$ *is bounded by a constant* $C > 0$ *independent of* $k \geq 0$ *and* $\ell \geq -k$. *Then there exists a constant* $K > 0$ *such that for any* $n \geq 1$, $\xi \in T_xM$ *and* $\eta \in T_{F^nM}M$,

$$K^{-1}\mathcal{J}_n(x) \leq (\det_{F^nx} \int_0^n DF^t(DF^t)^*dt)^{\frac{1}{2}} \leq Kn^{\upsilon^0}\mathcal{J}_n(x), \qquad (2.42)$$

$$\langle(\int_0^n DF^t(DF^t)^*dt)^{-1}\eta,\eta\rangle_{F^nx} \leq K\langle\eta,\eta\rangle, \qquad (2.43)$$

$$\langle(\int_0^n DF^{t-n}(DF^{t-n})^*dt)^{-1}\xi,\xi\rangle_x \leq K\langle\xi,\xi\rangle \qquad (2.44)$$

where $\upsilon^0 = \dim H^0_x$ *and* $\mathcal{J}_n(x)$ *is the same as in (2.7).*

Proof. Denote $A = \int_0^1 DF^t(DF^t)^*dt$ then

$$\int_0^n DF^t(DF^t)^*dt = \sum_{k=0}^{n-1} DF^kA(DF^k)^*. \qquad (2.45)$$

Put $\hat{H}^s_z = (H^0_z \oplus H^u_x)^\perp$, $\hat{H}^0_z = (H^s_z \oplus H^z_u)^\perp$, and $\hat{H}^u_z = (H^s_z \oplus H^0_z)^\perp$ where $L^\perp$ means the orthogonal complement of L and $z = F^kx$, $k = 0,1,\ldots$. Then $T_zM = \hat{H}^s_z \oplus \hat{H}^0_z \oplus \hat{H}^u_z$ and

$$(DF)^*\hat{H}^s_{Fz} = \hat{H}^s_z,\ (DF)^*\hat{H}^0_{Fz} = \hat{H}^0_z,\ (DF)^*\hat{H}^u_{Fz} = \hat{H}^u_z \qquad (2.46)$$

for all $z = F^kx$, $k = 0,1,\ldots$. Indeed, check, for instance, the first equality in (2.46). Let $\zeta \in \hat{H}^s_{Fz}$ and

$\eta \in H_z^0 \oplus H_z^u$ then $\langle (DF)^* \zeta, \eta\rangle_z = \langle \zeta, DF\eta\rangle_{Fz} = 0$ since $DF\eta \in H_{Fz}^0 \oplus H_{Fz}^u$. Since $(DF)^*$ is an isomorphism of corresponding linear spaces this proves the first relation in (2.46). Others are the same. We claim that $\hat{H}_z^s$, $\hat{H}_z^0$, $\hat{H}_z^u$, $z = F^k x$ have the same properties with respect to $(DF)^*$ as H_z^s, H_z^0, H_z^u have with respect to DF, namely for some $\tilde{C} > 0$ and all $k, \ell \geq 0$,

$$\|(DF^\ell)^* \xi\|_{F^k x} \leq \tilde{C} e^{-\gamma\ell} \|\xi\|_{F^{k+\ell}x} \quad \text{if} \quad \xi \in \hat{H}^s_{F^{k+\ell}x}, \tag{2.47}$$

$$\|(DF^{-\ell})^* \eta\|_{F^{k+\ell}x} \leq \tilde{C} e^{-\gamma\ell} \|\eta\|_{F^k x} \quad \text{if} \quad \eta \in \hat{H}^u_{F^k x}, \tag{2.48}$$

and for any $m \geq -k$,

$$\|(DF^m)^* \zeta\|_{F^k x} \leq \tilde{C} \|\zeta\|_{F^{k+m}x} \quad \text{if} \quad \zeta \in \hat{H}^0_{F^{k+m}x}. \tag{2.49}$$

Indeed, check, for instance, (2.47). If $\xi \in \hat{H}^s_{F^{k+\ell}x} = (H^0_{F^{k+\ell}x} + H^u_{F^{k+\ell}x})^\perp$ then by (2.20),

$$\|(DF^\ell)^*\xi\|^2_{F^k x} = \langle (DF^\ell)^*\xi, (DF^\ell)^*\xi\rangle_{F^k x} \tag{2.50}$$

$$= \langle \xi, DF^\ell (DF^\ell)^* (\xi^s + \xi^0 + \xi^u)\rangle_{F^{k+\ell}x}$$

$$= \langle \xi, DF^\ell (DF^\ell)^* \xi^s\rangle_{F^{k+\ell}x} \leq \tilde{K}\gamma^{-1} e^{-\gamma\ell} \|(DF^\ell)^*\xi^s\|$$

$$\leq \tilde{C} e^{-\gamma\ell} \|(DF^\ell)^*\xi\|$$

where $\xi = \xi^s + \xi^0 + \xi^u$ is the unique representation with $\xi^s \in H_z^s$, $\xi^0 \in H_z^0$, $\xi^u \in H_z^u$, $z = F^{k+\ell}x$, and constants $\tilde{K}, \tilde{C} > 0$ depend only on angles between the subspaces H_z^s, H_z^0, and H_z^u. To prove (2.48) and (2.49) we use (2.6) and the boundedness of $\|DF^m|_{H^0_{F^k x}}\|$ for all $k \geq 0$ and $m \geq -k$. Now the proof of (2.43) and (2.44) follows easily from (2.45)-(2.49), taking into account that A in (2.45) is a positive definite matrix. The assertion (2.42) follows from the same relations noticing that the cosine between spaces H^u_z and H^0_z, and also between $\hat{H}^u_z$ and $\hat{H}^0_z$ is bounded away from zero. The determinant in question is the coefficient with which the operator $\int_0^n DF^t(DF^t)^* dt$ increases the volume. Then this operator increases the volume on $\hat{H}^u_{F^n x}$ by $(\mathcal{J}_n(x))^2$, it may increase the volume on $\hat{H}^0_{F^n x}$ by at most Cn^{ν^0}, and the increase of the volume on $\hat{H}^s_{F^n x}$ is by a constant sandwiched between two positive constants. All this is a standard linear algebra and we leave the details to the reader or refer to Section 4 of Kifer [Ki6]. □

In fact we shall need in Section 2.7 the following lower bound

$$r^\epsilon_x(t,\xi,\eta) \geq (2\pi\epsilon^2)^{-n/2} e^{-2K} K^{-1} (\mathcal{J}_n(x))^{-1} n^{-\nu^0} \tag{2.51}$$

which follows from (2.41)-(2.44) provided $\xi \in T_x M$, $\eta \in T_{F^n x} M$, $\|\xi\| \leq \epsilon$ and $\|\eta\| \leq \epsilon$. It suffices to use in Section 2.7 the upper bound (2.7) which holds in more general situations than

$$r^\epsilon_x(n,\xi,\eta) \leq (2\pi\epsilon^2)^{-n/2} K (\mathcal{J}_n(x))^{-1} \tag{2.52}$$

which follows from (2.41) and (2.42).

Remark 2.1. In all results of this section the subspaces $\tilde{H}_{F^k x}$, $H^0_{F^k x}$, and $H^s_{F^k x}$ may be empty so that we shall have $T_{F^k x}M = H^u_{F^k x}$. This will be important for applications to expanding transformations.

2.3. Hyperbolic and expanding transformations.

In this section we recall definitions and review certain results concerning hyperbolic and expanding transformations needed in the subsequent exposition. In some points we only outline main arguments or principal ideas of proofs referring the reader for details to Shub [Sh] and other papers.

We shall start with the shadowing property for expanding transformations. A C^2-endomorphism F of a compact Riemannian manifold M is called expanding if there exists $\gamma > 0$ such that

$$\|DF^n\xi\| \geq \gamma e^{\gamma n}\|\xi\| \quad \text{for all} \quad \xi \in T_xM,\ x \in M,\ n \geq 0 \qquad (3.1)$$

where DF denotes the differential of the map F.

Proposition 3.1. *The shadowing property holds true for any expanding map* F *of a compact Riemannian manifold* M.

Proof. Choose an integer $n_0 > 0$ so that $q = \gamma e^{\gamma n_0} > 1$ and put $G = F^{n_0}$. Let $x_0, x_1, \ldots, x_n$ be a δ-pseudo-orbit of F. Then $y_0 = x_0$, $y_1 = x_{n_0}$, $y_2 = x_{2n_0}, \ldots, y_k = x_{kn_0}$ with k = integral part of $(\frac{n}{n_0})$ is a $\tilde{C}n_0\delta$-pseudo-orbit of G, where $\tilde{C} = (\|DF\|+1)^{n_0}$, $\|DF\| = \sup_{x\in M}\|DF\|_x$, and $\|DF\|_x = \sup_{\xi\in T_xM,\,\|\xi\|=1}\|DF\xi\|$. It is easy to see that if $R = \tilde{C}n_0(q-1)^{-1}$ and $\delta > 0$ is small enough then one of the preimages for the map G of the closed ball $\overline{U_{R\delta}(y_{i+1})}$ is contained in $\overline{U_{R\delta}(y_i)}$, $i = 0,1,\ldots,k-1$, i.e., if $V_i = \overline{U_{R\delta}(y_i)} \cap G^{-1}\overline{U_{R\delta}(y_{i+1})}$ then $GV_i = \overline{U_{R\delta}(y_{i+1})}$. We obtain the following picture.

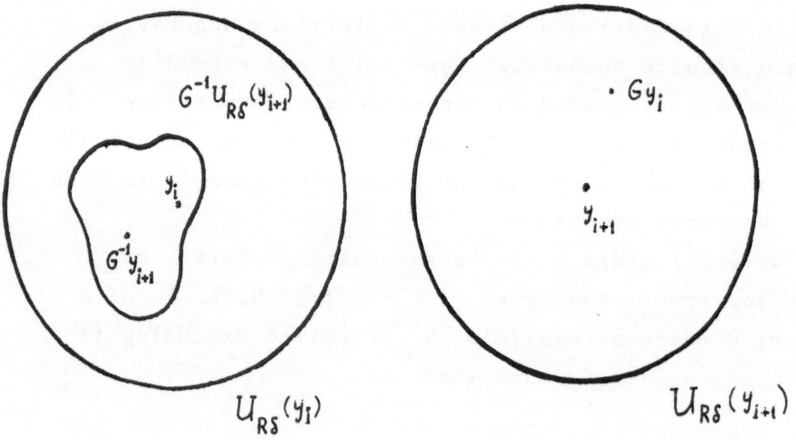

Figure 3.1

Thus the intersection $\bigcap_{i=0}^{k} G^{-i}\overline{U_{R\delta}(y_i)}$ is not empty and any point z from this intersection satisfies $G^i z \in \overline{U_{R\delta}(y_i)}$ for all $i = 0,1,\ldots,k$. By the definition of $\tilde{C}$ this implies also that $F^j z \in \overline{U_{\tilde{C}n_0 R\delta}(x_i)}$ for all $j = 0,1,\ldots,n$, and so the shadowing takes place with $V = M$, $\beta = 0$, and $C = \tilde{C}n_0 R$. □

Next, we shall discuss the shadowing property for different types of diffeomorphisms with hyperbolicity conditions whose simplest case was considered in Example 1.5. The proof in the general case is an extension of the argument we had in Example 1.5.

Definition 3.1. A compact set $\Lambda \subset M$ invariant with respect to a diffeomorphism F (i.e., $F\Lambda = \Lambda$) of a M is called hyperbolic if there is a Hölder continuous splitting $T_\Lambda M = H^s \oplus H^u$ of the tangent bundle TM of M restricted to Λ which is invariant with respect to the differential DF of F, $DFH^s = H^s$, $DFH^u = H^u$, and such that for some constant $\gamma > 0$ one has

$$\|DF^n\xi\| \leq \gamma^{-1}e^{-\gamma n}\|\xi\| \quad \text{if} \quad \xi \in H^s,\ n \geq 0 \tag{3.2}$$

and

$$\|DF^{-n}\eta\| \leq \gamma^{-1}e^{-\gamma n}\|\eta\| \quad \text{if} \quad \eta \in H^u,\ n \geq 0. \tag{3.3}$$

The subbundles H^s and H^u are called stable and unstable, respectively.

Remark 3.1. It is easy to see that the hyperbolicity of Λ does not depend on the choice of the metric (see Shub [Sh], p.20). Furthermore, if we have a DF-invariant splitting satisfying (3.2) and (3.3) then it will be automatically Hölder continuous (see, for instance, Appendix to Brin and Kifer [BK]).

Remark 3.2. A compact set Λ in a locally compact space always has an open neighborhood $U \supset \Lambda$ whose closure is compact. We shall always restrict our study to compact neighborhoods of hyperbolic sets and attractors (without mentioning this each time) though we are reluctant to assume the compactness of M since hyperbolic sets and attractors appear naturally in noncompact spaces, as well.

A diffeomorphism F of M is called Anosov if the whole M is a hyperbolic set for F. A simplest example of this situation is Example 1.5. An example of a non-Anosov hyperbolic set is Smale's Horseshoe (see, for instance, Shub [Sh]).

Let Λ be a hyperbolic set and $\zeta \in T_\Lambda M$. Then there is a unique representation $\zeta = \zeta^s + \zeta^u$ with $\zeta^s \in H^s$ and $\zeta^u \in H^u$. Define the new norm by

$$|||\zeta|||^2 = \sum_{n=0}^{\infty} (\|DF^n \zeta^s\|^2 + \|DF^{-n}\zeta^u\|^2). \tag{3.4}$$

It is easy to see that for some positive $\lambda < 1$,

$$|||DF\zeta^s||| \leq \lambda |||\zeta^s||| \text{ and } |||DF^{-1}\zeta^u||| < \lambda |||\zeta^u||| \tag{3.5}$$

(see Shub [Sh], chapter 4). The norm $|||\cdot|||$ may not be smooth since, in general, H^s and H^u need only be continuous subbundles (in fact, Hölder continuous). However, one can approximate $|||\cdot|||$ by a smooth norm, which we shall denote by the same sign $|||\cdot|||$, so that (3.5) will remain true with, maybe, a little bit bigger λ but still less than one. This norm generates a Riemannian metric, called an adapted metric, which can be extended smoothly into the whole M. An adapted metric essentially simplifies many arguments since it enables one to employ the hyperbolicity already for one step as in Example 1.5 and we avoid using an iterate of F as in Proposition 3.1.

Let Λ be a hyperbolic set for a diffeomorphism F of M. For any $x \in \Lambda$ define

$$\begin{aligned} W^s(x) &= \{y \in M : \mathrm{dist}(F^n x, F^n y) \to 0 \text{ as } n \to \infty\}, \\ W^s_\rho(x) &= \{y \in W^s(x) : \mathrm{dist}(F^n x, F^n y) \leq \rho \text{ for all } n \geq 0\}, \end{aligned} \tag{3.6}$$

$$\begin{aligned} W^u(x) &= \{z \in M : \mathrm{dist}(F^{-n} x, F^{-n} z) \to 0 \text{ as } n \to \infty\}, \\ W^u_\rho(x) &= \{z \in W^u(x) : \mathrm{dist}(F^{-n} x, F^{-n} z) \leq \rho \text{ for all } n \geq 0\}. \end{aligned} \tag{3.7}$$

Proposition 3.2. *Let* Λ *be a hyperbolic set of a* C^2 *diffeomorphism* F *furnished with an adapted metric. Then there is* $\rho > 0$ *small enough such that for every* $x \in \Lambda$ *the sets* $W^s_\rho(x)$ *and* $W^u_\rho(x)$ *(called local stable and unstable manifolds at* x*) are embedded disks tangent at* x *to* H^s_x *and* H^u_x*, respectively, and having the same dimension as* H^s_x *and* H^u_x*. The stable and unstable disks are as smooth as* F*, they depend continuously on* $x \in \Lambda$ *and satisfy*

$$d(F^n x, F^n y) \le \lambda^n d(x,y) \text{ for all } y \in W^s_\rho(x),\ n \ge 0 \quad (3.8)$$

and

$$d(F^{-n} x, F^{-n} z) \le \lambda^n d(x,z) \text{ for all } z \in W^u_\rho(x),\ n \ge 0, \quad (3.9)$$

where $0 < \lambda < 1$ *and* $d(\cdot,\cdot)$ *is an adapted metric extended into the whole* M*. The global stable and unstable manifolds* $W^s(x)$ *and* $W^u(x)$ *are immersed submanifolds of* M *diffeomorphic Euclidean spaces of corresponding dimensions.*

Proof can be found in Shub [Sh], Chapter 6. Remark that W^s and W^u are integral manifolds of subbundles H^s and H^u but usual methods of integration of subbundles of the tangent bundle such as the Frobenius theorem cannot be used here since H^s_x and H^u_x are not smooth in x, in general. The proof is based on the observation that under the action of DF the angle between vectors close to H^u and H^u decreases. Thus if we shall take a family of small smooth disks K_x, $x \in \Lambda$ embedded into M whose tangent vectors are close to H^u_x then under the iterates of F the tangent vectors of K_x approach H^u. One derives from here that $U_\rho(x) \cap F^n K_{F^{-n}x}$ tends as $n \to \infty$

to a limit in an appropriate Banach space. This limit will be the desired local unstable manifold. One obtains local stable manifolds in the same way by applying F^{-1} in place of F. The global stable and unstable manifolds are obtained by $W^s(x) = \bigcup_{n\geq 0} F^{-n}W^s_\rho(F^n x)$ and $W^u(x) = \bigcup_{n\geq 0} F^n W^u_\rho(F^{-n}x)$. □

Suppose that x and y belong to a hyperbolic set Λ and $\mathrm{dist}(x,y)$ is small enough. Since $W^s_\rho(x)$ and $W^u_\rho(y)$ are transverse and have complementary dimensions then $W^s_\rho(x) \cap W^u_\rho(y)$ is a single point denoted by $[x,y]$.

Definition 3.2. A hyperbolic set Λ is said to have a local product structure if there is $\delta > 0$ such that $[x,y] \in \Lambda$ whenever $x,y \in \Lambda$ and $\mathrm{dist}(x,y) \leq \delta$.

Remark 3.3. A hyperbolic set Λ is said to be locally maximal if there is an open set $U \supset \Lambda$ such that any F-invariant set $\tilde{\Lambda}$ satisfying $\Lambda \subset \tilde{\Lambda} \subset U$ coincides with Λ, equivalently $\bigcap_{-\infty<n<\infty} F^n U = \Lambda$. It turns out that this more transparent property is equivalent to a local product structure of Λ. Indeed, if Λ is a locally maximal hyperbolic set and points $x,y \in \Lambda$ are close to each other, then $z = [x,y]$, $\mathrm{dist}(F^n z, F^n x) \to 0$ and $\mathrm{dist}(F^{-n}z, F^{-n}y) \to 0$ as $n \to \infty$. Thus the whole orbit of the point z stays in a small neighborhood of Λ, and so z must belong to the maximal invariant set in this neighborhood which is Λ. To prove in the other direction, suppose that Λ has a local product structure but in any neighborhood of Λ there is a bigger than Λ invariant set $\tilde{\Lambda}$. The orbit of any point $v \in \tilde{\Lambda}\setminus\Lambda$ stays close to Λ, by Proposition 3.3 below F has the shadowing property on Λ, and so by Proposition 3.5 the orbit of v stays close to the orbit of a point $x \in \Lambda$ for positive time and it stays close to the orbit of another point $y \in \Lambda$ for negative time. From this one derives that $v \in W^s(x)$ and $v \in W^u(y)$, and so we must have $v = [x,y] \in \Lambda$.

Next, we shall establish the shadowing for hyperbolic sets (see also Shub [Sh], Chapter 8).

Proposition 3.3. *Let* Λ *be a hyperbolic set for a diffeomorphism* F *of a manifold* M. *Then there exists an open set* $U_\Lambda \supset \Lambda$ *such that* F *has the shadowing property on* U_Λ *with* $\beta = 0$. *If* Λ *has a local product structure then* F *has the shadowing property on* Λ *itself, i.e., if a* δ*-pseudo-orbit* $\omega = (z_0,\ldots,z_n)$ *stays entirely in* Λ *then a point* y_ω *in (1.18) can be chosen to belong* Λ, *as well. If* Λ *is an attractor then any open set satisfying (I.4.2) and having a compact closure will do.*

Proof. The hyperbolic DF-invariant splitting $T_\Lambda M = H^s \oplus H^u$ on Λ and the adapted metric can be extended to a continuous slitting $T_U M = \tilde{H}^s + \tilde{H}^u$ and smooth metric on some neighborhood U of Λ, so that

$$|||DF\xi||| \le \lambda|||\xi||| \text{ and } |||DF^{-1}\eta||| \le \lambda|||\eta||| \tag{3.10}$$

$$\text{for } \xi \in \tilde{H}^s,\ \eta \in \tilde{H}^u$$

where $0 < \lambda < 1$ (see Shub [Sh], Chapter 7). For each $x \in U$ take a cross product of ρ-disks in $\tilde{H}^s$ and $\tilde{H}^u$ in the $|||\cdot|||$-norm. Exponentiating these down into M one obtains some boxes $V_x(\rho)$. Then it follows easily that one can find $\delta_0 > 0$ small enough and $C > 0$ big enough such that if $U \subset \{y{:}\mathrm{dist}(y,\Lambda) \le \delta_0\}$ and $\delta \le \delta_0$ then for any δ-pseudo-orbit $\omega = (z_0,\ldots,z_n)$ staying entirely in U the image $FV_{z_{k-1}}(C\delta)$ stretches across the box $V_{z_k}(C\delta)$ in the unstable direction and does not intersect the boundary of $V_{z_k}(C\delta)$ in the stable direction. This yields (similarly to Example 1.5) that the intersection $\Gamma = \bigcap_{k=0}^{n} F^k V_{z_{n-k}}(C\delta)$ is not empty and any point of $F^{-n}\Gamma$ can be taken as the point y_ω in (1.18). If Λ has a local product structure and $z_0, \ldots, z_n \in \Lambda$ then one can show that $\Gamma \cap \Lambda \neq \emptyset$, and so in this case y_ω can be chosen to belong to Λ. Let now Λ be an attractor and U_Λ be an open set satisfying (I.4.2). Suppose that U is a neighborhood of Λ for which one can apply the arguments above. There exists $N > 0$ such that $F^N \bar{U}_\Lambda \subset U$. Then it is easy to see that any δ-pseudo-orbit $\omega = (z_0, \ldots, z_n)$ with $\delta > 0$ small enough will stay in U except maybe for N first points $z_0, \ldots, z_{N-1}$. Now we apply the first part of Proposition 1.4 to $z_N, \ldots, z_n$ obtaining some constant C in (1.18). Then (1.18) will be satisfied for the whole δ-pseudo-orbit with the constant C $\max_{-N \le n \le N} \sup_{x \in U_\Lambda} \|DF^n\|_x$ where

$$\|DF^n\|_x = \sup_{\xi \in T_x M, \|\xi\|=1} \|DF^n \xi\| \tag{3.11}$$

completing the proof. □

In all cases above we had the shadowing property with $\beta = 0$. Suppose now that F^t, $t \in (-\infty, \infty)$ is a flow generated by a smooth vector field $B(x)$ on M according to the ordinary differential equation (1.11).

Definition 3.3. A compact F^t-invariant $(-\infty < t < \infty)$ set $\Lambda \subset M$ is called hyperbolic with respect to the flow F^t if there is a DF^t-invariant Hölder continuous splitting $T_\Lambda M = H^s \oplus H^0 \oplus H^u$ of the tangent bundle TM restricted to Λ so that H^0 is the one-dimensional subbundle generated by the vector field B and for some constant $\gamma > 0$ the inequalities (3.2) and (3.3) hold true. Again, the subbundles H^s and H^u are called stable and unstable, respectively, and Remark 3.1 remains true.

Similarly to a diffeomorphism case, for any x from a hyperbolic set for a C^2 flow F^t one obtains stable and unstable manifolds $W^s(x)$, $W^s_\rho(x)$ and $W^u(x)$, $W^u_\rho(x)$ defined by (3.6)-(3.7) and satisfying (3.8)-(3.9). Suppose that x and y belong to a hyperbolic set Λ and $\text{dist}(x,y)$ is small enough. Since $W^s_\rho(x)$ and $W^u_\rho(x)$ are transverse both to each other and to the flow direction and the sum of all dimensions equal to the dimension of M then there exists a unique $t(x,y)$ small in absolute value such that $W^s_\rho(F^{t(x,y)}x) \cap W^u_\rho(y)$ is a single point denoted by $[x,y]$. Now we repeat Definition 3.2 to obtain the notion of a local product structure on Λ. Similarly to Remark 3.3 one calls a hyperbolic set Λ locally maximal if there is no bigger F-invariant set in a small neighborhood of Λ. Again, one can show that Λ has a local product structure if and only if it is locally maximal.

Proposition 3.4. *Let Λ be a hyperbolic set for a C^2 flow F^t of a manifold M. Then there exists an open set $U_\Lambda \supset \Lambda$ such that the diffeomorphism $F = F^1$ (called the time-one map of the flow) has the shadowing property on U_Λ with $\beta = 1$. If Λ has a local product structure then F has the shadowing property on Λ itself, i.e., if $\omega = (z_0,\dots,z_n)$ is a δ-pseudo-orbit for F staying entirely in Λ then a point y_ω in (1.18) can be chosen to belong to Λ. If Λ is an attractor then any open set with a compact closure satisfying (I.4.2) will do.*

Proof proceeds in the same way as in Proposition 3.3 but the boxes $V_x(C\delta)$ must be extended in the flow direction, as well, so that we shall obtain a nonempty intersection

$$\Gamma = \bigcap_{k=0}^{n} F^k \Big[\bigcup_{t:|t|\le C\delta(n-k+1)} V_{z_{n-k}}(C\delta) \Big]$$

(see Kifer [Ki6], Lemma 3.2). Then any point of $F^{-n}\Gamma$ can be taken as y_ω in (1.18). □

Different versions of the shadowing property can be established in more general circumstances (see Hirsch, Pugh and Shub [HPS]). It is not difficult to extend the argument of Propositions 3.3 and 3.4 in order to prove the shadowing in the form (1.18) near a compact set Λ such that the tangent bundle TM has a DF-invariant Hölder continuous splitting over Λ, $T_\Lambda M = H^s \oplus H^0 \oplus H^u$ with H^s and H^u satisfying (3.2) and (3.3), respectively, and the restriction of DF on H^0 satisfies $\|DF^n|_{H^0}\| \le C|n|^\sigma$ for some constants $C, \sigma > 0$ and any integer $n = 0, \pm 1, \pm 2, \ldots$. In this situation we called in [Ki10] the set Λ strongly partially hyperbolic. If $\sigma = 0$ we obtain a direct generalization of time-one maps of hyperbolic flows. This case occurs, for instance, when one considers an algebraic automorphism of the m-dimensional torus given by an m×m matrix with integer entries and the determinant equal to one, and having some of eigenvalues (but not all) equal to one in absolute value. Another example of this kind is the frame flow on negatively curved manifolds whose random perturbations were studied in [Ki1].

The shadowing property will enable us to study random perturbations X_n^ε only along tube neighborhoods of true orbits of iterates F^n. But this study can be accomplished adequately only for orbits staying entirely in the set Λ where we have a true invariant hyperbolic structure. The following implication of the shadowing property helps to overcome this difficulty.

Proposition 3.5. *Let* Λ *be a locally maximal hyperbolic set for a* C^2*-diffeomorphism* F *or* C^2*-flow* F^t *(and then* $F = F^1$*). Then there exist constants* $\tilde{C}, \tilde{\rho} > 0$ *such that for any* $n = 1,2,\ldots$ *and a point* $x \in U_\Lambda$ *satisfying* $\max_{0 \le k \le n} \operatorname{dist}(F^k x, \Lambda) \le \rho$ *with* $\rho \le \tilde{\rho}$ *one can find* $y \in \Lambda$ *so that*

$$\max_{0 \le k \le n} \operatorname{dist}(F^k x, F^k y) \le \tilde{C}\rho. \tag{3.12}$$

Proof (cf. Lemma 3.1 in Kifer [Ki6] and Corollary 3.1 in Kifer [Ki10]). Suppose first that Λ is a hyperbolic set for a diffeomorphism F. There exist points $z_k \in \Lambda$ satisfying $\operatorname{dist}(z_k, F^k x) \le \rho$ for all $k = 0,1,\ldots,n$. We can assume that $\Gamma = \{y : \operatorname{dist}(y,\Lambda) \le \tilde{\rho}\}$ is compact. Since Λ is F-invariant then $\operatorname{dist}(Fz_k, z_{k+1}) \le \tilde{\tilde{C}}\rho$ where $\tilde{\tilde{C}} = 1 + \sup_{x \in U_\Lambda} \|DF\|_x$ and $\|DF\|_x$ is defined by (3.11). Thus $\omega = (z_0, \ldots, z_n)$ is a $\tilde{\tilde{C}}\rho$-pseudo-orbit and by Proposition 3.3 it can be shadowed in the sense of (1.18) with $\beta = 0$ by the true orbit of a point $y = y_\omega \in \Lambda$ proving (3.12) with $\tilde{C} = C\tilde{\tilde{C}}$. In the flow case one chooses first points $\tilde{z}_k \in \Lambda$ satisfying $\operatorname{dist}(\tilde{z}_k, F^k x) \le \rho$ and then adjusts them by a small shift along orbits of F^t obtaining new points $z_k \in \Lambda$ which form a $K\rho$-pseudo-orbit such that it can be shadowed by the orbit of a point $y = y_\omega \in \Lambda$ satisfying (1.18) with $\beta = 0$. □

Let x belong to a hyperbolic set Λ. Denote by $\mathcal{J}(x)$ the absolute value of the Jacobian of the linear map $DF : H_x^u \to H_{Fx}^u$ restricted to the unstable subbundle where we use inner products induced by the Riemannian metric which determines $\mathcal{J}(x)$ uniquely. Clearly,

$\mathcal{J}_n(x) = \mathcal{J}(x)\mathcal{J}(Fx)\cdots\mathcal{J}(F^{n-1}x)$ is the absolute value of the Jacobian of the linear map $DF^n:H_x^u \to H_{F^nx}^u$. Remark that since the subbundle H^u is Hölder continuous then $\mathcal{J}(x)$ is a Hölder continuous function of $x \in \Lambda$. Define also, $d_n(x,y) = \max\{\text{dist}(F^kx,F^ky), 0 \le |k| \le |n| \text{ and } kn \ge 0\}$ and $K_\rho(x,n) = \{y: d_n(x,y) \le \rho\}$.

Proposition 3.6. *Let* Λ *be a hyperbolic set of a* C^2 *diffeomorphism* F. *Then there exist* $\tilde{\rho}$, C_ρ, $C > 0$ *such that for any positive* $\rho \le \tilde{\rho}$, $n \ge 0$ *and* $x \in \Lambda$,

$$C_\rho^{-1} \le m(K_\rho(x,n))\mathcal{J}_n(x) \le C_\rho, \tag{3.13}$$

where m *is the Riemannian volume, and for each* $y \in \Lambda \cap K_\rho(x,n)$,

$$C^{-1} \le \mathcal{J}_n(x)(\mathcal{J}_n(y))^{-1} \le C. \tag{3.14}$$

If Λ *is a hyperbolic set for a flow* F^t *and* $F = F^1$ *then* (3.13) *remains true and* (3.14) *must be replaced by*

$$C^{-1} \le \mathcal{J}_n(F^ux)(\mathcal{J}_n(y))^{-1} \le C \tag{3.15}$$

with $|u| \le c\rho$, *where* $c \ge 0$ *depends only on* Λ.

Proof can be found in Appendix to Bowen and Ruelle [BR] (remark that (3.13) is known since [BR] as the volume lemma). We shall outline the argument. If Λ is a hyperbolic set for a diffeomorphism F, $x \in \Lambda$, and $y \in K_\rho(x,n)$ then

$$\begin{aligned} &\text{dist}(F^kx,F^ky) \\ &\le \tilde{C}e^{-\sigma\min(k,n-k)}\max(\text{dist}(x,y),\text{dist}(F^nx,F^ny)) \\ &\le \tilde{C}e^{-\sigma k}\max(\text{dist}(x,y),\text{dist}(F^nx,F^ny)) \end{aligned} \tag{3.16}$$

for some constants $\tilde{C}, \sigma > 0$ independent of x, y, and n. Indeed, by the transversality of stable and unstable directions we can consider the orbit $F^k y$, $k = 0,1,\dots,n$ by means of "projections" on stable and unstable directions at corresponding points $F^k x$, $k = 0,\dots,n$. Since $y \in K_\rho(x,n)$ this will lead to a conclusion that y belongs to an exponentially in n narrow strip along $W^s(x)$ and $F^n y$ belongs to an exponentially in n narrow strip along $W^u(F^n x)$, and so when both k and $(n-k)$ are large $\mathrm{dist}(F^k x, F^k y)$ will be exponentially in $\min(k,n-k)$ small. The following picture illustrates this argument for the automorphism of the torus $F = \begin{pmatrix} 2 & 1 \\ 1 & 1 \end{pmatrix}$ considered in Example 1.5. Here y belongs to a narrow strip along the stable direction at x, $F^n y$ belongs to a narrow strip along the unstable direction at $F^n x$, and $F^k y$ belongs to a rectangle centered at $F^k x$ and small in any direction.

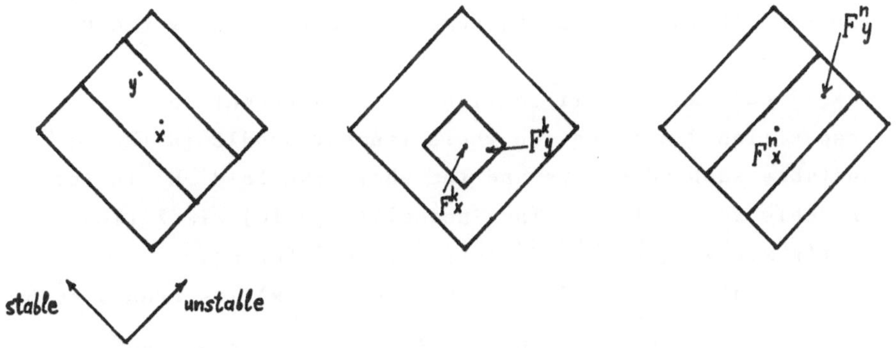

Figure 3.2

Suppose $\varphi(x)$ is Hölder continuous and $\varphi_n(z) = \varphi(z)\varphi(Fz)\cdots\varphi(F^{n-1}z)$ then (3.16) yields (3.14). If Λ is a hyperbolic set for a flow F^t and $F = F^1$ then the left hand side of (3.16) must be replaced by $\operatorname{dist}(F^{k+u}x, F^k y)$ making an adjustment in the flow direction by some u of order ρ. This gives (3.15). To obtain (3.13) we remark that $K_\rho(x,n) = \bigcap_{k=0}^{n} F^{-k}\overline{U_\rho(F^k x)}$, and so $K_\rho(x,n)$ has the size of order ρ in the stable (and flow) direction and its volume has the same order as the induced volume of the piece $A = W^u_{\hat{C}\rho}(x) \cap K_\rho(x,n)$ $= F^{-n}(W^u_{\hat{C}\rho}(F^n x) \cap \overline{U_\rho(F^n x)})$ of the unstable submanifold for some constant $\hat{C} > 0$. By (3.14) we conclude that $\operatorname{vol}(F^n A)$ $= \operatorname{vol}(W^u_{\hat{C}\rho}(F^n x) \cap \overline{U_\rho(F^n x)})$ is both of order $\varphi_n(x)\operatorname{vol} A$ and of order a constant depending on ρ, yielding (3.13).

□

Proposition 3.7. *Let* $F: M \to M$ *be a* C^2 *expanding map then (3.13) and (3.14) remain true for any* $x, y \in M$ *and* $n \geq 0$.

Proof is a simplification of the argument in Proposition 3.6 since the whole tangent bundle is the unstable subbundle here and the whole manifold M is the unstable submanifold. The inequality (3.16) simplifies to $\operatorname{dist}(F^k x, F^k y) \leq \tilde{C}e^{-\sigma(n-k)}\operatorname{dist}(F^n x, F^n y)$ for any $y \in K_\rho(x,n)$, and (3.14) follows since $\varphi(x)$ is even smooth here. If ρ is small then $K_\rho(x,n)$ is a connected component of the intersection $\overline{U_\rho(x)} \cap F^{-n}\overline{U_\rho(F^n x)}$ which contains the point x, implying (3.13).

□

A hyperbolic set which is also an attractor, i.e., it satisfies (I.4.2), will be called a hyperbolic attractor.

From (I.4.2) it follows that a hyperbolic attractor is always a locally maximal hyperbolic set, (see Remark 3.3), and so it has a local product structure.

Proposition 3.8. *Let* Λ *be a hyperbolic attractor of a* C^2 *diffeomorphism* $F:M \to M$. *Then*
(a) *For any* $x \in \Lambda$, $W^u(x) \subset \Lambda$ *and* $\text{int}\, \cup\{W^s_\rho(y): y \in W^u_\rho(x)\}$ *is an open neighborhood of* x *in* M.
(b) $W^s(\Lambda) = \bigcup_{y\in\Lambda} W^s(y) \supset U_\Lambda$ *where* $U_\Lambda \supset \Lambda$ *is an open set satisfying (I.4.2), and for any such set* U_Λ *there exist constants* $C, \sigma > 0$ *so that for each* $z \in U_\Lambda$ *one can find* $y \in \Lambda$ *satisfying* $z \in W^s_{C\text{dist}(z,\Lambda)}(y)$, *and for all* $n \geq 0$,

$$\text{dist}(F^n y, F^n z) \leq Ce^{-\sigma n}\text{dist}(y,z). \tag{3.17}$$

(c) *If* Λ *is a hyperbolic attractor for a* C^2 *flow* F^t, *and* $F = F^1$ *then for any* $x \in \Lambda$, $W^u(x) \subset \Lambda$ *and*

$$\bigcup_{|t|<\rho} F^t \text{int}\, \cup\{W^s_\rho(y): y \in W^u_\rho(x)\}$$

is an open neighborhood of x *in* M. *The assertion (b) remains true as stated.*

Proof. (a) Let U_Λ be a neighborhood of Λ satisfying (I.4.2). If $y \in W^u(x)$ then for all n bigger than some $n_0 \geq 0$, $F^{-n}y \in U_\Lambda$, i.e., $y \in \bigcap_{n>n_0} F^n U_\Lambda = \Lambda$. Hence if $x \in \Lambda$ then $W^s(y)$ exists for any $y \in W^u_\rho(x)$. Thus the set $\cup\{W^s_\rho(y): y \in W^u_\rho(x)\}$ is correctly defined and its interior is an open neighborhood of x since W^s and W^u are transverse and have complementary dimensions. The assertion (b) follows from (a), (3.8), and the compactness of Λ. The proof for the flow case is the same. □

The following result is called an absolute continuity of the stable foliation.

Proposition 3.9. *Let Λ be a hyperbolic attractor for a C^2 diffeomorphism F. Then there exist constants $C, \tilde{\rho} > 0$ such that if $0 \leq \rho \leq \tilde{\rho}$, $x, y \in \Lambda$, and Borel sets $E_1 \subset W^u_\rho(x)$, $E_2 \subset W^u_\rho(y)$ satisfy $E_2 = \bigcup_{z \in E_1} [z, y]$, with $[\cdot, \cdot]$ introduced before Definition 3.2, then*

$$C^{-1} \leq \frac{m^u(E_1)}{m^u(E_2)} \leq C \tag{3.18}$$

where m^u is the volume on unstable subbundles induced by the Riemanian metric. The same is true if Λ is a hyperbolic attractor for a C^2 flow F^t.

Proof can be found in Appendix A of Ruelle [Ru1]. We shall exhibit the argument. It suffices to obtain (3.18) for E_1 being an arbitrarily small ball $W^u_\delta(z) \subset W^u_\rho(x)$ for some $z \in W^u_\rho(x)$ and $\delta > 0$. Take $n > 0$ very big so that $F^n E_1$ and $F^n E_2$ will be extremely close to each other. Pick up $\eta > 0$ so small that $F^n W^u_{2\eta}(v) \subset W^u_\rho(F^n v)$ for any $v \in E_1$. Choose a maximal collection of points $v_i \in E_1$ with mutual distances bigger than 2η. Then $\bigcup_i W^u_{2\eta}(v_i) \supset E_1$ and the sets $W^u_\eta(v_i)$ are disjoint. Since $E_1 = W^u_\delta(z)$ and its boundary has zero m^u-volume then only tiny proportion of balls $W^u_{2\eta}(v_i)$ intersects the boundary of E_1 provided η is small enough. Thus we shall arrive at

$$m^u(E_1) \leq \sum_i m^u(W^u_{2\eta}(v_i)) \leq 3^\upsilon m^u(E_1) \tag{3.19}$$

where υ is the dimension of M. Now it is easy to see that (3.18) will follow if we show that

$$C^{-1} \le m^u(Q_i)/m^u(\tilde{Q}_i) \le C \tag{3.20}$$

for both $Q_i = W^u_\eta(v_i)$ and $Q_i = W^u_{2\eta}(v_i)$ where $\tilde{Q}_i = \bigcup_{z\in Q_i} [z,y]$. By the choice of n and η, $F^nQ_i \subset W^u_\rho(F^nv_i)$, and $F^n\tilde{Q}_i$ being extremely close to F^nQ_i can be obtained from F^nQ_i by "almost parallel displacement" along stable manifolds. This implies that the ratio $m^u(F^nQ_i)/m^u(F^n\tilde{Q}_i)$ is close to one. In view of (3.14) both ratios $m^u(F^nQ_i)/\mathcal{I}_n(v_i)m^u(Q_i)$ and $m^u(F^n\tilde{Q}_i)/\mathcal{I}_n([v_i,y])m^u(\tilde{Q}_i)$ are sandwiched between two positive constants. These together with (3.14) lead to (3.20) and finally prove (3.18). The proof in the flow case is similar but first we consider $E_1(t) = \bigcup_{0\le s\le t} F^sE_1$ for small t and $E_2(t) = \bigcup_{z\in E_1(t)} \varphi(z,y)$ where $\varphi(z,y)$ is the unique point of the intersection $W^s_\rho(z)$ and $W^u_\rho(F^\tau y)$ for some τ, $|\tau| \le \rho$. Then again $F^nE_1(t)$ and $F^nE_2(t)$ will be extremely close to each other and we can repeat the above argument to derive that if $\tilde{m}^u$ is the induced volume on unstable leafs $\mathcal{L}^u(z) = \bigcup_{-\infty<t<\infty} F^tW^u(z)$ then the ratio $\tilde{m}^u(E_1(t))/\tilde{m}^u(E_2(t))$ is sandwiched between two positive constants. On the other hand, it is easy to see that $\tilde{m}^u(E_1(t))$ is of order $tm^u(E_1)$ and $\tilde{m}^u(E_2(t))$ is of order $tm^u(E_2)$ implying the desired assertion. □

Remark 3.4. Clearly, Propositions 3.8 and 3.9 are not relevant in the case of expanding transformations since in this case the whole M is the unstable manifold of any point.

Proposition 3.10. *Let Λ be a hyperbolic attractor of a C^2 diffeomorphism F. There exist constants $C, \epsilon_0, \rho > 0$ such that if $x, y \in \Lambda$, $0 < \epsilon \leq \epsilon_0$, $n \geq (\log \epsilon)^2$, and the intersection $W^u_\epsilon(x) \cap F^{-n}W^s_\rho(y)$ consists of points $\{z_k\}$ then*

$$\sum_k (\mathcal{J}_n(z_k))^{-1} \leq C\epsilon^{\upsilon^u} \tag{3.21}$$

where $\upsilon^u = \dim W^u(x)$. If Λ is a hyperbolic attractor of a C^2 flow F^t and $F = F^1$ then the same is true if we shall take points $\{x_k\}$ of the intersection $W^u_\epsilon(x) \cap F^{-n}(\bigcup_{|t|\leq\rho} F^t W^s_\rho(y))$.

Proof. First, remark that the intersection $W^u_\epsilon(x) \cap F^{-n}W^s_\rho(y)$ consists of a finite number of points which all belong to Λ since $W^u_\epsilon(x)$ and $F^{-n}W^s_\rho(y)$ are uniformly transverse and $W^u(x) \subset \Lambda$. For any pair of points $v_1, v_2 \in \Lambda$ which are sufficiently close to each other and any sets of small diameter $E \subset W^u_\rho(v_1)$, $D = W^s_\rho(v_2) \cap \Lambda$ we denote $[E,D] = \{[v,w] : v \in E, w \in D\}$. By Proposition 3.8(a), $[E,D] \subset \Lambda$. Put $A = [W^u_\rho(y), W^s_{2\rho}(y) \cap \Lambda]$ where $\rho > 0$ is small but fixed. If ϵ is small enough and $n \geq (\log \epsilon)^2$ then connected components of the intersection $W^u_{2\epsilon}(x) \cap F^{-n}A$ are extremely small pieces of $W^u_{2\epsilon}(x)$, each of these pieces does not contain more than one point among $\{z_k\}$, those pieces which do contain points $\{z_k\}$ are disjoint both from the boundary of $W^u_{2\epsilon}(x)$ and from the set $[W^u_\rho(y), \partial W^s_{2\rho}(y) \cap \Lambda]$ where ∂ denotes the boundary. Employing (3.14) we conclude that the m^u-volume of a piece containing a point z_k is of order $m^u(W^u_\rho(y))(\mathcal{J}_n(z_k))^{-1}$. Thus

$$\tilde{C}\epsilon^{\nu^u} \geq m^u(W^u_{2\epsilon}(x)) \geq m^u(W^u_{2\epsilon}(x) \cap F^{-n}\Lambda) \tag{3.22}$$

$$\geq \tilde{C}^{-1} m^u(W^u_\rho(y)) \sum_k (\mathcal{J}_n(z_k))^{-1}$$

for some constant $\tilde{C} > 0$ depending only on Λ. This yields (3.21). In order to prove (3.21) for the flow case we put $A = [W^u_\rho(y), \bigcup_{|t| \leq 2\rho} F^t(W^s_{2\rho}(y) \cap \Lambda)]$, take the intersection $W^u_{2\epsilon}(x) \cap F^{-n}A$, and argue in the same way a above. □

Proposition 3.11. *Let* F *be a* C^2 *expanding transformation of a compact Riemannian manifold* M. *Then there exist constants* $C, \epsilon_0, \rho > 0$ *such that if* $x, y \in M$, $0 < \epsilon < \epsilon_0$, $n \geq (\log \epsilon)^2$, *and points* $z_1, \dots, z_k$ *are chosen so that one point is taken from each connected component of the intersection* $U_\epsilon(x) \cap F^{-n}U_\rho(y)$, *one has*

$$\sum_{1 \leq i \leq k} (\mathcal{J}_n(z_i))^{-1} \leq C\epsilon^\nu \tag{3.23}$$

where $\nu = \dim M$.

Proof proceeds in the same way as in Proposition 3.10 by a simplification of the above proof, and so we leave the details to the reader. □

2.4. Limiting measures.

In this section we shall establish absolute continuity in unstable directions of weak limits or random perturbations near a hyperbolic attractor.

Let Λ be a hyperbolic attractor for a C^2 diffeomorphism F or for a C^2 flow F^t, and then

$F = F^1$. We shall assume that random perturbations X_n^ϵ and the attractor Λ satisfy Assumptions 1.1 and 1.2 with an open set $U_\Lambda \supset \Lambda$ satisfying (I.4.2). Let $\Gamma \subset M$ be a Borel set. According to Lemma 1.1, our mistake will be at most $\gamma_1(\epsilon, n)m(\Gamma) = Cn\epsilon^{-2\nu}m(\Gamma)e^{-\frac{\alpha\delta}{2\epsilon}}$ if we shall compute the n-step transition probabilities $P^\epsilon(n, x, \Gamma)$, $x \in U_\Lambda$ of Markov chains X_k^ϵ taking into account only paths of X_k^ϵ which are δ-pseudo-orbits starting at x and ending in Γ. Since we assume that $\overline{FU_\Lambda} \subset U_\Lambda$ then if $\delta > 0$ is small enough any δ-pseudo-orbit $\omega = (x, y_1, \ldots, y_n)$ has all points in U_Λ. Thus by Propositions 3.3 and 3.4 one can find $z^\omega \in U_\Lambda$ such that

$$\text{dist}(y_k, F^k z^\omega) \le Cn\delta, \; y_0 = x, \; k = 0, \ldots, n, \tag{4.1}$$

and so by (3.17),

$$\text{dist}(y, \Lambda) \le \tilde{C}(n\delta + e^{-\sigma n}) \equiv \gamma_2(\delta, n). \tag{4.2}$$

Set

$$n = n(\epsilon) = [(\log \epsilon)^2 + 1], \; \delta = \delta(\epsilon) = \epsilon^{1-\beta} \tag{4.3}$$

where $\beta < 1$, $\beta > 0$ will be chosen later and $[a]$ denotes the integral part of a. Then if ϵ is small enough then

$$\gamma_1(\epsilon, n) \le \exp(-\epsilon^{-\beta/2}) \text{ and } \gamma_2(\delta, n) \le \epsilon^{1-3/2\beta}. \tag{4.4}$$

According to Proposition 3.8(b) for any $\rho > 0$ small enough, $W^s_{C\rho}(\Lambda) \supset U_\rho(\Lambda)$ where $W^s_r(\Lambda) = \bigcup_{z \in \Lambda} W^s_r(z)$, $U_\rho(\Lambda) = \bigcup_{z \in \Lambda} U_\rho(z)$, and $C > 0$ is independent of ρ. Thus by (1.14) and (4.2)-(4.4),

$$P^{\epsilon}(n,x,W^{s}_{C\gamma_2(\epsilon,n)}(\Lambda)) \geq P^{\epsilon}(n,x,U_{\gamma_2(\epsilon,n)}) \tag{4.5}$$

$$\geq 1-\gamma_1(\epsilon,n)m(U_\Lambda) \geq 1-m(U_\Lambda)\exp(-\epsilon^{-\beta/2}).$$

If μ^{ϵ} is an invariant measure of the Markov chain X^{ϵ}_n with the support in U_Λ then by (1.1), and (4.5),

$$\mu^{\epsilon}(W^{s}_{C\gamma_2(\epsilon,n)}(\Lambda)) \geq \mu^{\epsilon}(U_{\gamma_2(\epsilon,n)}(\Lambda)) \tag{4.6}$$

$$\geq 1-m(U_\Lambda)\exp(-\epsilon^{-\beta/2}),$$

i.e., except for a very small mass the measure μ^{ϵ} concentrates in a $\epsilon^{1-2\beta}$-neighborhood of Λ. In particular, of course, if

$$\mu^{\epsilon_i} \xrightarrow{w} \mu \quad \text{then} \quad \text{supp}\,\mu \subset \Lambda. \tag{4.7}$$

If Borel sets $E \subset E^{u}_{\rho}(v_1)$, and $D \subset W^{s}_{\rho}(v_2) \cap \Lambda$, $v_1, v_2 \in \Lambda$ of small diameter are closures of their interiors in $W^{u}(v_1)$ and $W^{s}(v_2) \cap \Lambda$, respectively, then $[E,D] = \{[v,w]: v \in E, w \in D\}$ is called a rectangle. By the definition and Proposition 3.8(a), $[E,D] \subset \Lambda$. The main goal of this section is to establish the following.

Theorem 4.1. *There exist* ρ_0, $C > 0$ *such that for any rectangle* $[E,D]$ *with* E *and* D *having diameters not exceeding* ρ_0 *and for any probability measure* μ *on* Λ *which is a weak limit as* $\epsilon_i \to 0$ *of a sequence* μ^{ϵ_i} *of invariant measures of Markov chains* $X^{\epsilon_i}_n$ *satisfying Assumptions 1.1 and 1.2 one has*

$$\mu([E,D]) \le Cm^u(E) \tag{4.8}$$

in the case that Λ *is a hyperbolic attractor for a diffeomorphism* F, *where, recall,* m^u *is the induced volume on unstable manifolds. If* Λ *is a hyperbolic attractor for a flow* F^t, $F = F^1$ *then under the same conditions*

$$\mu\Big(\bigcup_{t:|t|\le\rho} F^t[E,D]\Big) \le Cm^u(E). \tag{4.9}$$

Proof. It suffices to prove (4.8) and (4.9) for $E = W^u_\eta(z)$ and $D = W^s_\rho(z) \cap \Lambda$ for all $z \in \Lambda$ and any $\eta,\rho > 0$ small enough. Choose points $v_i \in E$, $i = 1,\dots,k_\epsilon$ so that

$$E \subset \bigcup_i W^u_\epsilon(v_i) \quad \text{and} \quad \sum_i m^u(W^u_\epsilon(v_i)) \le 3^\nu m^u(E). \tag{4.10}$$

This can be done in the same way as in (3.19) if ϵ is small enough by taking a maximal collection of points in E with mutual distances bigger than ϵ noticing that $W^u_{\epsilon/2}(v_i)$ are disjoint, and only tiny proportion of sets $W^u_\epsilon(v_i)$ intersects the boundary ∂E since its m^u-volume is zero.

For any set $\Gamma \subset \Lambda$ we denote $W^s_\rho(\Gamma) = \bigcup_{y\in\Gamma} W^s_\rho(y)$. For the simultaneous treatment of the diffeomorphism and the flow case define also $\tilde{W}^s_\rho(\Gamma) = W^s_\rho(\Gamma)$ in the diffeomorphism case and $\tilde{W}^s_\rho(\Gamma) = \bigcup_{t:|t|\le\rho} W^s_\rho(\Gamma)$ in the flow case. Denote

$$I^\epsilon_1(\delta,n,x,\Gamma) = P^\epsilon_x\{\mathrm{dist}(FX^\epsilon_i, X^\epsilon_{i+1}) < \delta$$
$$\text{for all } i = 0,\dots,n-1 \text{ and } X^\epsilon_n \in \Gamma\}.$$

Then by (1.14), (4.3)-(4.5) and (4.9), for ϵ small enough

$$P^{\epsilon}(n(\epsilon),x,\tilde{W}^s_{\rho}(E)) \tag{4.11}$$

$$\le I_1^{\epsilon}(\delta(\epsilon),n(\epsilon),x,\tilde{W}^s_{\rho}(E) \cap W^s_{\epsilon^{1-2\beta}}(\Lambda)) + \exp(-\epsilon^{-\beta/3})$$

$$\le \sum_{1\le i\le k_{\epsilon}} I_1^{\epsilon}(\delta(\epsilon),n(\epsilon),x,A_i^{\epsilon}) + \exp(-\epsilon^{-\beta/3})$$

where $A_i^{\epsilon} = \tilde{W}^s_{\rho}(E_i^{\epsilon}) \cap W^s_{\epsilon^{1-2\beta}}(\Lambda)$, $E_i^{\epsilon} = W^u_{\epsilon}(v_i)$, and $x \in W^s_{\epsilon^{1-2\beta}}(\Lambda)$, say $x \in W^s_{\epsilon^{1-2\beta}}(\tilde{x})$ with $\tilde{x} \in \Lambda$. Let $G_j^{\epsilon} = \overline{W^u_{j\epsilon}(\tilde{x})\backslash W^u_{(j-1)\epsilon}(\tilde{x})}$ for $j = 1,2,\dots,[\epsilon^{-3\beta}]+1$ where $[\cdot]$ denotes the integral part. Let the intersection $G_j^{\epsilon} \cap F^{-n(\epsilon)}\tilde{W}^s_{2\rho}(v_i)$ consists of points z_{ijk}, $k = 1,\dots,k^{(ij)}$. If $\omega = (x,y_1,\dots,y_n)$ is a $\delta(\epsilon)$-pseudo-orbit such that $x \in W^s_{\epsilon^{1-2\beta}}(\tilde{x})$, $\tilde{x} \in \Lambda$ and $y_n \in A_i^{\epsilon}$ then by (4.4) and Proposition 3.3 (or by Proposition 3.4 in the flow case) there exists a point y^{ω} satisfying

$$\operatorname{dist}(y_{\ell},F^{\ell}y^{\omega}) \le \epsilon^{1-2\beta} \quad \text{for all } \ell = 0,1,\dots,n(\epsilon) \tag{4.12}$$

provided ϵ is small enough, where $y_0 = x$. In particular,

$$y^{\omega} \in \tilde{W}^s_{\epsilon^{1-3\beta}}(W^u_{\epsilon^{1-3\beta}}(\tilde{x}))$$

and

$$F^{n(\epsilon)}y^{\omega} \in \tilde{W}^s_{\epsilon^{1-3\beta}}(W^u_{\epsilon^{1-3\beta}}(A_i^{\epsilon}))$$

where we denote $W^u_r(\Gamma) = \bigcup_{v\in\Gamma} W^u_r(v)$ for any $\Gamma \subset \Lambda$. But then, clearly, one can find i,j,k with $j \le [\epsilon^{-3\beta}]+1$ such that $\operatorname{dist}(F^{\ell}y^{\omega},F^{\ell}z_{ijk}) \le \frac{1}{2}\epsilon^{1-4\beta}$ for all

$\ell = 0,1,\ldots,n(\epsilon)$ with z_{ijk} introduced above and ϵ small enough. Hence by (4.12),

$$\operatorname{dist}(y_\ell, F^\ell z_{ijk}) \le \epsilon^{1-4\beta} \quad \text{for all} \quad \ell = 0,\ldots,n(\epsilon). \tag{4.13}$$

Therefore

$$I_1^\epsilon(\delta(\epsilon), n(\epsilon), x, A_i^\epsilon) \tag{4.14}$$

$$\le \sum_{j \le [\epsilon^{-3\beta}]+1, k} I_2^\epsilon(\epsilon^{1-4\beta}, n(\epsilon), x, z_{ijk}, A_i^\epsilon)$$

where

$$I_2^\epsilon(\rho, n, x, z, \Gamma) = P_x^\epsilon\{\operatorname{dist}(X_\ell^\epsilon, F^\ell z) \le \rho \quad \text{for all}$$

$$\ell = 1,\ldots,n(\epsilon) \text{ and } X_n^\epsilon \in \Gamma\}$$

$$= \int_{U_\rho(Fz)} \cdots \int_{U_\rho(F^{n-1}z)} \int_{U_\rho(F^n z)\cap\Gamma} q_{Fx}^\epsilon(y_1) q_{Fy_1}^\epsilon(y_2)$$

$$\cdots q_{Fy_{n-1}}^\epsilon(y_n) dm(y_1)\cdots dm(y_n).$$

Let $4\beta < \alpha$ then by (1.7) for ϵ small enough and $z = z_{ijk}$,

$$I_2^\epsilon(\epsilon^{1-4\beta}, n, x, z, A_i^\epsilon) \tag{4.15}$$

$$\le (1+\epsilon^\alpha)^n \int_{U_{\epsilon^{1-4\beta}}(Fz)} \cdots \int_{U_{\epsilon^{1-4\beta}}(F^{n-1}z)} \int_{U_{\epsilon^{1-4\beta}}(F^n z)\cap A_i^\epsilon}$$

$$\times\ \epsilon^{-\upsilon} r_{Fx}\left(\tfrac{1}{\epsilon}\operatorname{Exp}_{Fx}^{-1} y_1\right) \epsilon^{-\upsilon} r_{Fy_1}\left(\tfrac{1}{\epsilon}\operatorname{Exp}_{Fy_1}^{-1} y_2\right)$$

$$\cdots \epsilon^{-\upsilon} r_{Fy_{n-1}}\left(\tfrac{1}{\epsilon}\operatorname{Exp}_{Fy_{n-1}}^{-1} y_n\right) dm(y_1)\cdots dm(y_n).$$

It is easy to see that there exists a constant $K > 0$ such that if $\text{dist}(Fy_{\ell-1}, F^{\ell}z) \le \epsilon^{1-4\beta}$, $\ell = 1,\dots,n = n(\epsilon)$ then

$$\|\text{Exp}^{-1}_{F^{\ell}z} y_{\ell} - DF\text{Exp}^{-1}_{F^{\ell-1}z} y_{\ell-1} - \pi_{Fy_{\ell-1}F^{\ell}z} \text{Exp}^{-1}_{Fy_{\ell-1}} y_{\ell}\| \tag{4.16}$$

$$\le K\epsilon^{2-8\beta}$$

for all $\ell = 1,\dots,n(\epsilon)$ where, recall, π_{vw} is the parallel displacement from T_vM to T_wM.

Set $\eta_{\ell} = \text{Exp}^{-1}_{F^{\ell}z} y_{\ell}$. Then by (4.16) and the property (1.9) of the functions $r_v(\xi)$ it follows that

$$r_{Fy_{\ell-1}}\left(\frac{1}{\epsilon}\text{Exp}^{-1}_{Fy_{\ell-1}} y_{\ell}\right) \le r_{F^{\ell}z}\left(\frac{1}{\epsilon}(\eta_{\ell} - DF\eta_{\ell-1})\right) \tag{4.17}$$

$$+ \epsilon^{1-9\beta} + \chi_{\partial^{\epsilon}_{\ell}}(y_{\ell}) r_{Fy_{\ell-1}}\left(\frac{1}{\epsilon}\text{Exp}^{-1}_{Fy_{\ell-1}} y_{\ell}\right)$$

where $\partial^{\epsilon}_{\ell} = \{\tilde{y} : \frac{1}{\epsilon}\text{Exp}^{-1}_{Fy_{\ell-1}} \tilde{y} \in \partial V^{+}_{Fy_{\ell-1}}(\epsilon^{1-9\beta})\}$, the domains V^{+}_{z} where defined in the condition (iii) of Assumption 1.1, and, as usual, χ_{Γ} denotes the indicator function of a set Γ.

Since the exponential mapping Exp_y is the diffeomorphism (of class C^2 in our case) of some neighborhood of zero in T_yM onto some neighborhood of y in M having the identity matrix as its Jacobian matrix at zero of T_yM, then

$$1 - \tilde{C}\,\mathrm{dist}(y,\tilde{y}) \le \frac{m(d\tilde{y})}{m_y(d\mathrm{Exp}_y^{-1}\tilde{y})} \le 1 + \tilde{C}\,\mathrm{dist}(y,\tilde{y}) \tag{4.18}$$

for some $\tilde{C} > 0$ independent of y, $\tilde{y}$, provided $\mathrm{dist}(y,\tilde{y})$ is small enough (all points we are talking about belong to a compact neighborhood of Λ), where, recall, m_y is the Riemannian volume in T_yM. From (4.18) and the condition (iii) in Assumption 1.1 it follows that

$$\int_{\partial_\ell^\epsilon \cap U_{\epsilon^{1-4\beta}}(F^\ell z)} \epsilon^{-\upsilon} r_{Fy_{\ell-1}}\left(\tfrac{1}{\epsilon}\mathrm{Exp}_{Fy_{\ell-1}}^{-1} y_\ell\right) dm(y_\ell) \tag{4.19}$$

$$\le 2C\epsilon^{1-9\beta}$$

provided $y_{\ell-1} \in U_{\epsilon^{1-4\beta}}(F^{\ell-1}z)$ where $C > 0$ is the same as in (1.8) and ϵ is small enough.

Since the manifolds $W^s(y)$ and $W^u(y)$, as well, as their tangent at $y \in \Lambda$ subspaces H_y^s and H_y^u are Hölder continuous in y (see Remark 3.1), then one can see that for any $y \in \Lambda$, $\tilde{y} \in U_y(\epsilon^{1-5\beta}) \cap \Lambda$, and vectors $\zeta, \eta \in \mathrm{Exp}_y^{-1}(W^s_{\epsilon^{1-5\beta}}(\tilde{y}))$ the angle between $\zeta-\eta$ and H_y^s is of order $\epsilon^{\gamma_0(1-5\beta)}$, where $\gamma_0 > 0$ is the corresponding Hölder exponent. Hence if $\zeta, \eta \in \mathrm{Exp}_y^{-1}(W^s_{\epsilon^{1-5\beta}}(\tilde{y}))$ then $\|\zeta^u-\eta^u\| \le \tilde{K}\epsilon^{(1-5\beta)(1+\gamma_0)}$ for some $\tilde{K} > 0$ independent of ϵ, ζ, η, y, $\tilde{y}$, where ζ^u and η^u are corresponding components in the unique decomposition $\xi = \xi^s+\xi^u$, $\xi^s \in H^s$ and $\xi^u \in H^u$ (or $\xi = \xi^s+\xi^0+\xi^u$, $\xi^0 \in H^0$, in the flow case). Take $\beta > 0$ so small that

$$(1-5\beta)(1+\gamma_0) > 1. \tag{4.20}$$

Then from the definition of A_i^ϵ it follows that there exists a constant $\hat{C} > 0$ independent of ϵ, z and ξ

such that

$$\Psi^u_\epsilon(\xi) \subset H^u_{F^n z}(\hat{C}\epsilon) \quad \text{for any} \quad \xi \in H^s_{F^n z} \tag{4.21}$$

(in the flow case we take $\xi \in H^s_{F^n z} + H^0_{F^n z}$) where $H^u_y(r)$ is an r-ball in H^u_y centered at zero, $\Psi^u_\epsilon(\xi) = \{\eta \in H^u_{F^n z} : \eta+\xi \in \Psi_\epsilon\}, \Psi_\epsilon = \mathrm{Exp}^{-1}_{F^n z}(A^\epsilon_i \cap U_{F^n z}(\epsilon^{1-4\beta}))$, and ϵ is small enough. Thus

$$\Psi_\epsilon \subset H^s_{F^n z}(1) + H^u_{F^n z}(\hat{C}\epsilon) \equiv H(\epsilon). \tag{4.22}$$

It is clear that for some $\hat{K} > 0$ independent of ϵ, $z \in \Lambda$ and $y \in U_\Lambda$,

$$\hat{K}^{-1}\epsilon^{\upsilon^u} \leq m^u(E^\epsilon_i) \leq \hat{K} m^u_{F^n z}(H^u_{F^n z}(\hat{C}\epsilon)) \leq \hat{K}^2 \epsilon^{\upsilon^u} \tag{4.23}$$

$$\leq \hat{K}^3 m^u(E^\epsilon_i)$$

and

$$\hat{K}^{-1}\epsilon^{\upsilon(1-4\beta)} \leq m(U_{\epsilon^{1-4\beta}}(y)) \leq \hat{K}\epsilon^{\upsilon(1-4\beta)}. \tag{4.24}$$

After these preparations we are ready to substitute in (4.15) the integrations in M by integrations in the tangent bundle and to employ Proposition 2.1 which will lead to the desired estimate. Thus from (2.2), (4.15), (4.17)-(4.19) and (4.21)-(4.24), substituting $\xi = \mathrm{Exp}^{-1}_z x$ and $\eta_\ell = \mathrm{Exp}^{-1}_{F^\ell z} y_\ell$ one obtains

$$I_2^{\epsilon}(\epsilon^{1-4\beta},n,x,z,A_i^{\epsilon}) \tag{4.25}$$

$$\le (1+\epsilon^{\alpha})^n(1+\tilde{C}\epsilon^{1-4\beta})^n\Big[I_3^{\epsilon}(\epsilon^{1-4\beta},n,\xi,z,F(\epsilon))$$

$$+ \sum_{1\le\tau\le n} ((\hat{K}\epsilon^{1-4\beta\upsilon-3\beta})^{\tau} + (2C\epsilon^{1-9\beta})^{\tau})$$

$$\times \prod_{\substack{1\le\kappa\le\tau \\ \ell_1<\ell_2<\cdots<\ell_\tau}} \sup_{\eta\in T_{F^{\ell_\kappa+1}z}(\epsilon^{1-4\beta})}$$

$$\times I_3^{\epsilon}(\epsilon^{1-4\beta},\ell_{\kappa+1}-\ell_\kappa-1,\eta,F^{\ell_\kappa+1}z,T_{F^{\ell_{\kappa+1}}z}(\epsilon^{1-4\beta}))\Big]$$

where $I_3^{\epsilon}(\rho,k,\zeta,v,\Psi) = \int_{T_{Fv}(\rho)} \cdots \int_{T_{F^{k-1}v}(\rho)} \int_{\Psi}$

$r_v^{\epsilon}(\zeta,\eta_1)r_v^{\epsilon}(\eta_1,\eta_2)\cdots r^{\epsilon}_{F^{k-1}v}(\eta_{k-1},\eta_k)dm_{Fv}(\eta_1)\cdots dm_{F^kv}(\eta_k)$

and $T_w(\rho) = \{\zeta \in T_wM: \|\zeta\| \le \rho\}$. Since we estimate the right hand side of (4.15) by means of (4.17) then the first I_3^{ϵ}-integral in the right hand side of (4.25) emerges when we substitute the first term of the right hand side of (4.17), and the complicated sum in (4.25) emerges when we substitute the remaining two terms from the right hand side of (4.17) (ℓ_κ, $\kappa = 1,\ldots,\tau$ are precisely places where we substitute these two terms). By (2.5), clearly,

$$I_3^{\epsilon}(\rho,k,\zeta,v,\Psi) \le R_v^{\epsilon}(k,\zeta,\Psi). \tag{4.26}$$

Applying this together with (2.7) and (2.8) to (4.25) we derive

$$I_2^\epsilon(\epsilon^{1-4\beta},n,x,z,A_i^\epsilon) \le (1+\epsilon^\alpha)^n(1+\tilde{C}\epsilon^{1-4\beta})^n \tag{4.27}$$

$$\times \{\tilde{C}\epsilon^{-\nu^u} m^u(E_i^\epsilon)(\mathcal{J}_n(z))^{-1}\exp(-\sigma\|\xi^u\|\epsilon^{-1})$$

$$+ (\mathcal{J}_n(z))^{-1} \sum_{1\le\tau\le n} \tilde{C}^\tau \epsilon^{\tau(1-4\beta\nu-9\beta-4\beta\nu^u)}\}$$

$$\le \tilde{K}(\mathcal{J}_n(z))^{-1}\epsilon^{-\nu^u} m^u(E_i^\epsilon)(\exp(-\sigma\|\xi^u\|\epsilon^{-1})+\epsilon^{(1-18\nu\beta)}),$$

provided

$$0 < \beta < (18\nu)^{-1}, \tag{4.28}$$

and ϵ is small enough, where $\tilde{C},\tilde{K} > 0$ are constants independent of x, z, i, ϵ, and n provided $n \ge (\log \epsilon)^2$.

By (4.11) and (4.14),

$$P^\epsilon(n(\epsilon),x,\tilde{W}_\rho^s(E)) \tag{4.29}$$

$$\le \exp(-\epsilon^{-\beta/3}) + \sum_{j\le[\epsilon^{-3\beta}]+1,k} I_2^\epsilon(\epsilon^{1-4\beta},n(\epsilon),x,z_{ijk},A_i^\epsilon).$$

From (3.14) (or (3.15) in the flow case) one has that

$$C^{-1} \le \mathcal{J}_n(z_{i_1jk})(\mathcal{J}_n(z_{i_2jk}))^{-1} \le C \tag{4.30}$$

for all $i_1,i_2 = 1,\dots,[\epsilon^{-3\beta}]+1$. Since the foliations W^s and W^u are uniformly transversal and continuous it follows from the definition of the points z_{ijk} that

$$\text{if } \xi = \mathrm{Exp}_{z_{ijk}}^{-1} x \text{ then } C_1^{-1} j\epsilon \le \|\xi^u\| \le C_1 j\epsilon \tag{4.31}$$

where $C_1 > 0$ is independent of x, i, j, k. Now

employing (4.10) one obtains from (4.27), (4.30) and (4.31) that

$$\sum_i I_2^{\epsilon}(\epsilon^{1-4\beta}, n(\epsilon), x, z_{ijk}, A_i^{\epsilon}) \tag{4.32}$$

$$\leq C_2 \epsilon^{-\upsilon^u} (\vartheta_{n(\epsilon)}(z_{i_1 jk}))^{-1} m^u(E)(e^{-\sigma j C_1^{-1}} + \epsilon^{(1-18\upsilon\beta)}),$$

for some $C_2 > 0$ independent of ϵ, x, i, j, k. It is clear that there exist points $v_1^{(j)}, \dots, v_{q_j}^{(j)} \in G_j^{\epsilon}$ such that $G_j^{\epsilon} \subset \bigcup_{1 \leq \ell < q_j} W_{\epsilon}^u(v_{\ell}^{(j)})$ where $q_j \leq C_3 j^{\upsilon^u - 1}$, and again $C_3 > 0$ depends only on Λ and F. Applying Proposition 3.10 to each $W_{\epsilon}^u(v_{\ell}^{(j)})$ separately one can see from (4.30) that

$$\epsilon^{-\upsilon^u} \sum_k (\vartheta_{n(\epsilon)}(z_{i_1 jk}))^{-1} \leq C_4 \tag{4.33}$$

for some $C_4 > 0$ independent of ϵ and k. Now (4.32) and (4.33) yield

$$\sum_{i,k} I_2^{\epsilon}(\epsilon^{1-4\beta}, n(\epsilon), x, z_{ijk}, A_i^{\epsilon}) \leq C_5 m^u(E) j^{\upsilon^u} \tag{4.34}$$

$$\leq C_5 m^u(E) j^{\upsilon^u} (e^{-\sigma j C_1^{-1}} + \epsilon^{(1-18\upsilon\beta)}).$$

Since $j \leq [\epsilon^{-3\beta}]+1$ in (4.29) one obtains from (4.29) and (4.34) that

$$P^{\epsilon}(n(\epsilon), x, \tilde{W}_{\rho}^s(E)) \leq C_6 m^u(E) + \exp(-\epsilon^{-\beta/3}) \tag{4.35}$$

assuming

$$0 < \beta < (25\nu)^{-1}. \tag{4.36}$$

From (1.1), (4.5) and (4.6) we obtain

$$\mu^{\epsilon}(\tilde{W}^s_{\rho}(E)) \leq \int_{W^s_{\epsilon^{1-2\beta}}(\Lambda)} d\mu^{\epsilon}(x) P^{\epsilon}(n(\epsilon), x, \tilde{W}^s_{\rho}(E)) \tag{4.37}$$

$$+ \exp(-\epsilon^{-\beta/3})$$

$$\leq C_6 m^u(E) + 2\exp(-\epsilon^{-\beta/3}).$$

Recall that $E = W^u_{\eta}(z)$. Let $E_k = W^u_{\eta+\frac{1}{k}}(z)$. Since $C_6 > 0$ does not depend on ρ and η provided they are sufficiently small then

$$\mu^{\epsilon}(\tilde{W}^s_{\rho+\frac{1}{k}}(E_k)) \leq C_6 m^u(E_k) + 2\exp(-\epsilon^{-\beta/3}). \tag{4.38}$$

Since $\overline{U_{\Lambda}}$ is compact one can find a subsequence $\epsilon_i \to 0$ such that

$$\mu^{\epsilon_i} \xrightarrow{w} \mu \quad \text{as} \quad \epsilon_i \to 0 \tag{4.39}$$

(and the following arguments remain true for any such subsequence). Then by (4.38),

$$\mu(\tilde{W}^s_{\rho}(E)) \leq \mu(\text{int } \tilde{W}^s_{\rho+\frac{1}{k}}(E_k)) \tag{4.40}$$

$$\leq \lim_{i\to\infty} \mu^{\epsilon_i}(\text{int } \tilde{W}^s_{\rho+\frac{1}{k}}(E_k)) \leq C_6 m^u(E_k)$$

where int denotes the interior. Letting $k \to \infty$ we have

$$\mu(\tilde{W}^s_\rho(E)) \le C_6 m^u(E_k). \tag{4.41}$$

In view of the uniform transversality of W^s and W^u there exists a constant $C_7 > 0$ so that if η is small enough then $\tilde{W}^s_\rho(E) \supset [E,D]$ if $D = W^s_{C_7^{-1}\rho}(z) \cap \Lambda$ and ρ is also sufficiently small (in the flow case $W^s_\rho(E) \supset \bigcup_{t:|t|\le C_7^{-1}\rho} F^t[E,D]$). The proof of Theorem 4.1 is complete. □

Theorem 4.2. *Let* F *be a* C^2 *expanding map of a compact Riemannian manifold* M. *There exists* $C > 0$ *such that for any Borel set* $E \subset M$ *and for any probability measure* μ *which is a weak limit of a sequence* μ^{ϵ_i} *of invariant measures of Markov chains* $X_n^{\epsilon_i}$ *as* $\epsilon_i \to 0$ *one has*

$$\mu(E) \le Cm(E). \tag{4.42}$$

Proof proceeds in the same way as above, just by disregarding all arguments concerning stable subbundles and stable submanifolds, and using Propositions 3.1 and 3.7. This is, actually, a simplification of the above proof and the details are left to the reader. □

Remark 4.1. The above absolute continuity results can be proved for some strongly partially hyperbolic dynamical systems discussed after Proposition 3.4. But the uniqueness of measures μ satisfying (4.8) or (4.9) or (4.42) is connected with certain questions of ergodic theory which will be addressed in the next section.

2.5. Sinai-Bowen-Ruelle's measures. Discussion.

In this section we shall review some results about the uniqueness of invariant measures absolutely continuous in the unstable direction which we obtained in Theorems 4.1 and 4.2. At the end we discuss other approaches to the problem.

We shall start with the following.

Proposition 5.1. *Let* Λ *be a hyperbolic attractor of a diffeomorphism* F *or of a flow* F^t *and then put* $F = F^1$. *If* μ *is an F-invariant probability measure on* Λ *satisfying (4.8) or (4.9) (in the flow case) then*

$$h_\mu(F) = \int_\Lambda \log \mathcal{J}(x) d\mu(x) \qquad (5.1)$$

where $h_\mu(F)$ *is the entropy defined in Section 1.2 and* $\mathcal{J}(x)$ *was introduced before Proposition 3.6.*

Proof. We shall outline the argument. For convenience drop in the definition of rectangles $[E,D]$ the requirement that E and D are closed sets. Since any intersection of rectangles is again a rectangle and a difference of two rectangles can be represented as a finite union of rectangles (i.e., rectangles form a semialgebra of sets) then we can easily arrive to a finite partition $\xi = (A_1,\ldots,A_\ell)$ consisting of rectangles of arbitrarily small diameter. By the Shannon-McMillan- Breiman theorem (see Proposition I.2.3) for μ-almost all $x \in \Lambda$ the limit $r(x) = \lim_{n\to\infty} \frac{1}{n} \log \mu(\xi_n(x))$ exists and

$$h_\mu(F,\xi) = -\int_\Lambda r(x) d\mu(x) \qquad (5.2)$$

where $\xi_n(x) = A_{i_0} \cap F^{-1}A_{i_1} \cap \cdots \cap F^{-(n-1)}A_{i_{n-1}}$ provided $F^k x \in A_{i_k}$ for $k = 0,1,\ldots,n-1$. In view of (3.13) and Theorem 4.1 it is easy to see that

$$\mu(\xi_n(x)) \leq C(\mathcal{J}_n(x))^{-1} \tag{5.3}$$

where $C > 0$ is independent of $x \in \Lambda$ and $n = 1,2,\ldots$. Since $\mathcal{J}_n(x) = \mathcal{J}(x)\mathcal{J}(Fx)\cdots\mathcal{J}(F^{n-1}x)$ then by the Birkhoff ergodic theorem (see, for instance, Walters [Wa], p.34) for μ-almost all x the limit

$$\lambda(x) = \lim_{n\to\infty} \frac{1}{n} \log \mathcal{J}_n(x) \tag{5.4}$$

exists and

$$\int \lambda(x)d\mu(x) = \int \log \mathcal{J}(x)d\mu(x). \tag{5.5}$$

Now (5.2)-(5.5) yield

$$h_\mu(F,\xi) \geq \int_\Lambda \log \mathcal{J}(x)d\mu(x). \tag{5.6}$$

On the other hand,

$$h_\mu(F) \leq \int_\Lambda \log \mathcal{J}(x)d\mu(x) \tag{5.7}$$

which is a partial case of the Ruelle inequality (see, for instance, Ledrappier [Le1], Theorem 2.2, Chapter II). Since $h_\mu(F) = \sup_\xi h_\mu(F,\xi)$ we obtain (5.1). □

Proposition 5.2. *Let* Λ *be a hyperbolic attractor for a diffeomorphism* F *or for a flow* F^t *(and then* $F = F^1$*) such that there exists an orbit of* F *or* F^t*, respectively, which is dense in* Λ*. Then the equality (5.1) determines an F-invariant measure* μ *uniquely and this measure will be denoted from now on by* μ_Λ*. The measure* μ_Λ *is ergodic (and even Bernoulli) and it has conditional measures on unstable submanifolds which are* *equivalent (and not only absolutely continuous) with respect to the Riemannian volume there.*

The proof of this result was obtained for the diffeomorphism case by Ruelle [Ru1] using the machinery of Markov partitions, and for the flow case the result was proved by Bowen and Ruelle [BR] employing the equilibrium states technique. They also showed that μ_Λ can be obtained in the following way. Let $\tilde{m}$ be an arbitrary probability measure having support in a small neighborhood of the hyperbolic attractor Λ and absolutely continuous with respect to the Riemannian volume. Define $\mu^{(n)}(\Gamma) = \tilde{m}(F^{-n}\Gamma)$ then $\mu^{(n)} \xrightarrow{w} \mu_\Lambda$ as $n \to \infty$. For Anosov diffeomorphisms the measures with the above properties were first introduced and studied by Sinai [Si1]. The methods of all these works have some relation to ideas of statistical mechanics (see Ruelle [Ru2]). By the above reasons the measure μ_Λ is called the Sinai-Bowen-Ruelle measure.

Now Theorem 4.1, Propositions 5.1 an 5.2 imply

Theorem 5.3. *In conditions of Proposition 5.2 the invariant measures μ^ε of Markov chains X_n^ε satisfying Assumptions 1.1 and 1.2 converge in the weak sense to μ_Λ as $\varepsilon \to 0$.*

For a C^2 expanding map F the uniqueness and the ergodicity of a measure satisfying (4.42) was proved by Krzyzewski and Szlenk [KS] (see also Szlenk [Sz], Theorem 5.5.4). Moreover, they proved that such a measure is equivalent to the Riemannian volume. This gives

Theorem 5.4. *In conditions of Theorem 4.2 the invariant measures μ^ε of random perturbations X_n^ε converge in the weak sense to an F-invariant measure which is ergodic and equivalent to the Riemannian volume.*

We proved Theorem 5.3 under Assumption 1.2 which restricts the study to one attractor. In fact, in Section

2.4 we needed only the time of order $(\log \epsilon)^2$ to ensure that the transition probabilities of X_n^ϵ for this time become in the unstable direction "almost" absolutely continuous with respect to the Riemanian volume. According to Corollary 1.1 it is very unlikely for the process X_n^ϵ to exit from a small neighborhood of an attractor during the time of order $(\log \epsilon)^2$. Suppose now that M is a compact manifold and Assumptions I.5.1, I.5.4, and II.1.1 are satisfied. Then according to Theorem I.5.4 and Corollary I.5.3 all weak limits as $\epsilon \to 0$ of invariant measures μ^ϵ of X_n^ϵ on M have support in the union of attractors. Moreover this support is being contained in $\bigcup_{i \in L_{min}} K_i$ where L_{min} was defined before Theorem I.5.4. These arguments lead to the following result.

Theorem 5.5. *In the above circumstances suppose that all attractors* K_i *with* $i \in L_{min}$ *are hyperbolic. If* μ *is any weak limit of measures* μ^ϵ *as* $\epsilon \to 0$ *then the restriction of* μ *on each* K_i, $i \in L_{min}$ *is proportional to the Sinai-Bowen-Ruelle measure* μ_{K_i}, *i.e.,*

$$\mu = \sum_{i \in L_{min}} \mu(K_i)\mu_{K_i}, \quad \sum_{i \in L_{min}} \mu(K_i) = 1, \ \mu(K_i) \geq 0.$$

Recall that the Oseledec multiplicative ergodic theorem asserts that if μ is an invariant probability measure of a diffeomorphism $F:M \to M$ such that $\int_M \log\|DF\|_x d\mu(x)$ exists then for μ-almost all x there exist numbers $\lambda_1(x) > \cdots > \lambda_{r(x)}(x)$ and a decomposition $T_xM = L_1(x) \oplus \cdots \oplus L_{r(x)}(x)$ of the tangent space at x such that for every nonzero tangent vector $v \in L_i(x)$,

$$\lim_{n \to \pm\infty} \frac{1}{n} \log\|DF^n v\| = \lambda_i(x). \tag{5.8}$$

The numbers $\lambda_i(x)$, $i = 1,\dots,r(x)$ are called the Lyapunov (characteristic) exponents of F at x; $\dim E_i(x)$ is called the multiplicity of $\lambda_i(x)$. The functions $r(x)$, $\lambda_i(x)$ and $\dim E_i(x)$ are F-invariant, and so they are constants μ-almost surely if μ is an ergodic measure.

Put

$$\lambda^+(x) = \sum_{1 \le i \le r(x)} \max(\lambda_i(x), 0) \dim E_i(x). \qquad (5.9)$$

The Ruelle inequality (see Ledrappier [Le1], Theorem 2.2 or Kifer [Ki9], Section 5.1) which we employed in Proposition 5.1 asserts that

$$h_\mu(F) \le \int_M \lambda^+(x) d\mu(x). \qquad (5.10)$$

It follows easily from the full version of the Oseledec theorem that if μ has its support in a hyperbolic set Λ of F or $F = F^1$ and Λ is a hyperbolic set of a flow F^t then

$$\int_\Lambda \lambda^+(x) d\mu(x) = \int_\Lambda \log \mathcal{J}(x) d\mu(x) \qquad (5.11)$$

since the sum of positive exponents precisely characterizes the volume growth in the unstable direction (and the rest is the ergodic theorem). Thus (5.10) implies (5.7). Furthermore the equality (5.1) is a partial case of the more general relation

$$h_\mu(F) = \int_M \lambda^+(x) d\mu(x) \qquad (5.12)$$

called the Pesin formula. It turns out that (5.12) is equivalent to the property that μ has absolutely continuous conditional measures on unstable submanifolds which exist for general not necessarily hyperbolic diffeomorphisms. This was proved by a very sophisticated technique in Ledrappier and Strelcyn [LS] (that the absolute continuity in unstable directions implies (5.12)), and in Ledrappier [Le2] and Ledrappier and Young [LeY] (in the other direction).

There is no hope to extend our method to this general situation to obtain that all limiting measures of random perturbations must be absolutely continuous in unstable directions. Moreover one needs additional assumptions to claim the existence or the uniqueness of such measures. They do exist in the case of partially hyperbolic attractors Λ (see Pesin and Sinai [PS]), i.e., when $T_\Lambda M = H^s \oplus H^0 \oplus H^u$ with H^s and H^u satisfying (3.2) and (3.3), and DF^n may contract and expand along an integrable subbundle H^0 slower than in H^s and H^u, respectively. Many steps of our proof can be carried out for what we called in the discussion after Proposition 3.4 strongly partially hyperbolic dynamical systems. In this situation we can arrive at the absolute continuity of limiting measures in unstable directions but the uniqueness of such measures can be proved only in special cases.

Remark 5.1. A natural case for an extension of our methods is the example of Katok [Ka] of a two dimensional Bernoulli diffeomorphism obtained from an automorphism of a torus by slowing down near a fixed point and, then, by stretching its neighborhood in order to make the Lebesgue measure to be invariant again. In this example the system lacks the hyperbolicity in a single point and outside of a small neighborhood of this point many nice properties such as shadowing are preserved. The main difficulty for the direct application of our method to this situation is the fact that random perturbations X_n^ε exit from, say, ε-neighborhood of the "bad" point (where nothing can be controlled) for the time of order ε^{-1}. This time is too big to apply our linearization procedure, for instance, $(1+\varepsilon^\alpha)^n$ in (4.15) will not be of order 1 then.

In view of the Ruelle inequality (5.7) (and (5.10)) the measure μ_Λ maximizes the expression

$$h_\mu(F) - \int_\Lambda \log \mathcal{J}(x)d\mu(x) \qquad (5.13)$$

and by (5.1) the maximum is zero when Λ is an attractor. By this reason and the similarity with the statistical mechanics μ_Λ is called an equilibrium state. In a more general case when Λ is hyperbolic set but not an attractor then the number

$$P(\Lambda) = \sup\{h_\mu(F)-\int_\Lambda \log \mathcal{J}(x)d\mu(x) : \mu \text{ is F-invariant} \qquad (5.14)$$

$$\text{and supp } \mu \subset \Lambda\}$$

is strictly negative (see Bowen and Ruelle [BR], Theorem 5.6). This supremum is, in fact, attained on some measure called again an equilibrium state, but we shall be interested in Section 2.7 in the number $P(\Lambda)$ itself called the topological pressure.

Next, we shall review other approaches to the study of limiting measures of random perturbations of hyperbolic diffeomorphisms due to Sinai [Si1] and Young [You]. Sinai's model explained in Example 1.2 concerns localized random perturbations of an Anosov diffeomorphism F. Markov chains X_n^ϵ with invariant measures μ^ϵ can be viewed as stationary, and so paths can be considered both for positive and negative time. Put $\omega = \{\ldots, X_{-n}^\epsilon, \ldots, X_0^\epsilon, \ldots, X_n^\epsilon, \ldots\}$ and let R^ϵ be the measure in the space Ω of such paths invariant with respect to the shift transformation θ. A version of the shadowing property implies that for each ω there exists a unique point $z = z(\omega) \in M$ such that $\text{dist}(F^k z(\omega), X_k^\epsilon(\omega)) \leq C\epsilon$, $k = 0, \pm 1, \ldots$. Thus we shall have a map $f:\Omega \to M$, $f(\omega) = z(\omega)$ such that $f\theta = Ff$ and f transforms R_ϵ into a measure η_ϵ on M invariant under F. Consider $\{f^{-1}(z), z \in M\}$ which is a measurable partition of Ω. Then R_ϵ will generate a system of conditional probabilities together with corresponding conditional processes which turn out to be nonhomogeneous in time ergodic Markov chains. Let $p_z^\epsilon(\cdot)$ be the conditional density of the distribution of $X_0^\epsilon(\omega)$ provided $\omega \in f^{-1}(z)$. Then it is easy to see that

$$\mu_\epsilon(\Gamma) = \int_M d\eta_\epsilon(z)\int_\Gamma p_z^\epsilon(y)dm(y). \qquad (5.15)$$

The density $p_z^\epsilon(y)$ is zero when $\mathrm{dist}(y,z) > C\epsilon$, and so (5.15) implies that the weak limits $\underset{\epsilon\to 0}{\text{w-lim}}\, \mu_\epsilon$ and $\underset{\epsilon\to 0}{\text{w-lim}}\, \eta_\epsilon$ coincide if at least one of them exist. Then Sinai [Si1] suggests to study the asymptotic behavior of η_ϵ as $\epsilon \to 0$ using the fact that this measure is F-invariant and representing it as a Gibbs measure (see [Si1]) constructed by means of the measure with maximal entropy and a rather complicated function g_ϵ. The main difficulty of this method is the proof of Hölder continuity of the function g_ϵ and the study of its behavior as $\epsilon \to 0$. As far as the author knows the complete proof of this argument never appeared. Besides this method has a severe restriction due to the fact that for the construction of the map f above all paths of random perturbations must be two-sided δ-pseudo-orbits, i.e., the perturbations must be localized. This eliminates Examples 1.3 and 1.4 and, actually, does not allow any continuous time model.

Recently, Young [You] suggested another model of random perturbations by means of random diffeomorphisms (see Kifer [Ki9]). Suppose that Λ is a hyperbolic attractor of a C^2 diffeomorphism F and $U_\Lambda \supset \Lambda$ is an open set satisfying (I.4.2). For $\alpha,\beta > 0$ put $\Omega_{\alpha,\beta}(F) = \{g \in C^1(\overline{U}_\Lambda) : d_{C^1}(F,g) \le \alpha$ and $L(Dg) \le \beta\}$ where $C^1(V)$ denotes the space of C^1 maps from V into V,

$d_{C^1}(\cdot,\cdot)$ is the C^1 distance between two maps, and $L(\cdot)$ is the Lipschitz constant of a map.

Theorem 5.6. *In the above circumstances let* $\alpha > 0$ *be sufficiently small and let* $\beta > L(DF)$. *For each* $\epsilon > 0$, *let* η_ϵ *be a Borel probability measure on* $\Omega_{\alpha,\beta}(F)$ *and let* μ_ϵ *be an invariant measure for the Markov chain* X_n^ϵ *with transition probabilities*

$$P^\epsilon(x,\Gamma) = \eta_\epsilon\{g : gx \in \Gamma\}. \tag{5.16}$$

If for every $\epsilon > 0$ *and* $x \in \overline{U}_\Lambda$, $P^\epsilon(x,\cdot)$ *is absolutely continuous with respect to the Riemannian volume on* M *then* μ_ϵ *weakly converges as* $\epsilon \to 0$ *to the Sinai-Bowen-Ruelle measure* μ_Λ.

For the proof we refer the reader to Young [You]. Remark that though this theorem requires only absolute continuity of transition probabilities and it does not need specific conditions like our Assumption 1.1, it is not known precisely when transition probabilities of Markov chains can be represented in the form (5.16) (see Section 1.1 of Kifer [Ki9]) which is equivalent to the representation

$$X_n^\epsilon = F_n^\epsilon \circ \cdots \circ F_1^\epsilon X_0^\epsilon \tag{5.17}$$

where $F_1^\epsilon, \ldots, F_n^\epsilon$ are mutually independent $\Omega_{\alpha,\beta}(F)$-valued random variables with the distribution η_ϵ. The representations (5.16)-(5.17) can be accomplished (see Kifer [Ki9], Section 5.2) for diffusion type random perturbations considered in Examples 1.3 and 1.4. But these models are out of the framework of Theorem 5.6 by essential reasons. The representation (5.17) for diffusion type perturbations involve diffeomorphisms which are far away from the unperturbed diffeomorphism F, and so they do not belong to $\Lambda_{\alpha,\beta}(F)$. This is a crucial point since the proof of Theorem 5.6 uses the deterministic persistence of the hyperbolic structure on Λ. Namely, it employs the fact that all C^1 close to F maps g also have hyperbolic attractors close to Λ (see Shub [Sh], Chapter 7) with stable and unstable directions close to those of F, and so an application of each such g actually smooths measures in unstable directions. Moreover the same reason does not allow one to extend this approach to situations where the hyperbolic structure is not stable under small (deterministic!) perturbation of a diffeomorphism as it happens to be, for instance, for partially hyperbolic or nonuniformly hyperbolic dynamical systems.

Remark 5.2. According to Sinai [Si2] an attractor Λ of an endomorphism $F:M \to M$ is called stochastic if there exists an open set $U \supset \Lambda$ such that for any absolutely continuous with respect to the Riemannian volume Borel probability measure μ having a support in U the sequence $\mu^{(n)}, \mu^{(n)}(\Gamma) = \mu(F^{-n}\Gamma)$, weakly converges to a probability measure μ_Λ whose support is Λ and which is independent of μ, F-invariant, and mixing with respect to the restriction of F to Λ. Let random perturbations X_n^ϵ be restricted to U in the same way as in Assumption 1.2. Since μ_Λ is mixing and $\sup \mu_\Lambda = \Lambda$ then F has a dense orbit in Λ. Thus a not very restrictive additional condition such as the left hand side of (I.4.10) or $r_x(0) > 0$, $x \in U$ in Assumption 1.1 will enable us to use Proposition I.1.8 to derive that there exists a unique invariant measure μ^ϵ of X_n^ϵ with a support in U. Let ρ be a metric on the space of Borel probability measures on U which generates the topology of weak convergence. Then by the triangle inequality

$$\rho(\mu_\Lambda, \mu^\epsilon) \le \rho(\mu_\Lambda, \mu^{(n)}) + \rho(\mu^{(n)}, \mu_\epsilon^{(n)}) + \rho(\mu_\epsilon^{(n)} \mu^\epsilon) \quad (5.18)$$

where $\mu_\varepsilon^{(n)}(\Gamma) = \int_U d\mu(x)P^\varepsilon(n,x,\Gamma)$. Clearly, for a fixed n, $\rho(\mu^{(n)},\mu_\varepsilon^{(n)}) \to 0$ as $\varepsilon \to 0$. Since $\rho(\mu_\Lambda,\mu^{(n)}) \to 0$ as $n \to \infty$ then if one can show that uniformly in $\varepsilon > 0$, $\rho(\mu_\varepsilon^{(n)},\mu^\varepsilon) \to 0$ as $n \to \infty$ then it follows $\rho(\mu_\Lambda,\mu^\varepsilon) \to 0$ as $\varepsilon \to 0$ which would be a nice result. Unfortunately, this last assertion seems to be rather difficult. Blank [Bl], Remark 3.1 thought that this must follow from the Doeblin condition as in Proposition I.1.8 and from the existence of a dense orbit of F on Λ. But, clearly, the number k appearing in Proposition I.1.8 will depend here on ε since in order to arrive to a bounded away from zero transition density one needs the number of steps at least of order $\log(\frac{1}{\varepsilon})$ which is the minimal time required for an ε-pseudo-orbit to escape at a fixed distance from the true orbit of an initial point. This means that Proposition I.1.8 may provide the convergence of $\rho(\mu_\varepsilon^{(n)},\mu^\varepsilon)$ to zero only when $n(\log \frac{1}{\varepsilon})^{-1} \to \infty$ which is not sufficient here.

2.6. Entropy via random perturbations.

In this section we shall show that in the case of a hyperbolic attractor Λ one can achieve equalities in (I.2.8) and (I.2.9) obtaining, therefore, the entropy $h_{\mu_\Lambda}(F)$ through certain entropies of random perturbations.

The main goal of this section is the proof of the following.

Theorem 6.1 (cf. Theorem I.2.5). *Let Λ be a hyperbolic attractor for a C^2-diffeomorphism F or for a C^2-flow F^t (and then we put $F = F^1$) such that there exists an orbit of F or F^t, respectively, which is dense in Λ. Suppose that random perturbations X_n^ε satisfy Assumptions 1.1 and 1.2, and let μ^ε denote an invariant measure of X_n^ε with a support in U_Λ which according to*

Theorem 5.3 *converges weakly to* μ_Λ *as* $\epsilon \to 0$. *Then there exists* $\rho_0 > 0$ *such that for any partition* $\Pi = (V_1, \dots, V_k)$ *of* M *into Borel sets satisfying*

$$\mu_\Lambda(\partial\Pi) = 0 \quad \text{and} \quad \max_{V_i \in \Pi_\rho(\Lambda)} \operatorname{diam} V_i \le \rho_0, \tag{6.1}$$

one has

$$\lim_{\epsilon \to 0} h^\epsilon(\theta, \zeta^\epsilon) = h_{\mu_\Lambda}(F) \tag{6.2}$$

where $\partial\Pi = \bigcup_{i=1}^{k} \partial V_i$, $\partial V_i = \overline{V}_i \setminus \operatorname{int} V_i$, $\Pi_\rho(\Lambda) = \{V_i : V_i \cap U_\rho(\Lambda) \ne \emptyset\}$, $U_\rho(\Lambda) = \{y : \operatorname{dist}(y, \Lambda) < \rho\}$, ζ^ϵ *is the partition of the sample space* Ω *into the sets* $\Gamma_j^\epsilon = \{\omega : X_0^\epsilon \in V_j\}$, $h_{\mu_\Lambda}(F)$ *is the entropy of* F *relative to the measure* μ_Λ, *and* $h^\epsilon(\theta, \zeta^\epsilon)$ *is the entropy of the shift transformation* θ *defined in Section* 1.2.

Proof. It will be convenient here to simplify the notations of Section 2.1. Put

$$\sigma^\epsilon(i_0, \dots, i_{n-1}) = \int_{V_{i_0}} d\mu^\epsilon(x) P_x^\epsilon\{X_1^\epsilon \in V_{i_1}, \dots, X_n^\epsilon \in V_{i_n}\}$$

and

$$H_n^\epsilon(\Pi) = -\sum_{i_0, \dots, i_{n-1}} \sigma^\epsilon(i_0, \dots, i_{n-1}) \log \sigma^\epsilon(i_0, \dots, i_{n-1}).$$

Then (6.2) can be written in the form

$$\lim_{\epsilon \to 0} \lim_{n \to \infty} \frac{1}{n} H_n^\epsilon(\Pi) = h_{\mu_\Lambda}(F). \tag{6.3}$$

Remark that according to Theorem I.2.5 we have always

$$\limsup_{\epsilon\to 0}\lim_{n\to\infty}\frac{1}{n}H_n^{\epsilon}(\Pi)\leq h_{\mu_\Lambda}(F). \tag{6.4}$$

We shall call a subset $\mathcal{E}\subset\Lambda$ (ρ,n)-separated if whenever $x,y\in\mathcal{E}$, $x\neq y$ it follows that $d_n(x,y)>\rho$, where d_n is the metric defined before Proposition 3.6. It is clear that if two different points y and z belong to a (ρ,n)-separated set $\mathcal{E}$, then the sets $K_{\rho/2}(y,n)$ and $K_{\rho/2}(z,n)$ defined before Proposition 3.6 are disjoint, and so $\mathcal{E}$ is a finite set and the cardinality of $\mathcal{E}$ does not exceed some constant depending just on ρ and n. Hence there exists a maximal (ρ,n)-separated set $\mathcal{E}(\rho,n)$ (not necessarily unique) in the sense that any set containing $\mathcal{E}(\rho,n)$ cannot be (ρ,n)-separated.

If $\mathcal{E}(\rho,n)$ is a maximal (ρ,n)-separated set of a hyperbolic attractor Λ then by Proposition 3.8(b),

$$\bigcup_{x\in\mathcal{E}(\rho,n)}K_{2\rho}(x,n)\supset\bigcup_{x\in\Lambda}K_\rho(x,n)\supset W_\rho^s(\Lambda)\supset U_{C_1^{-1}\rho}(\Lambda) \tag{6.5}$$

for some constant $C_1>0$ independent of $\rho>0$. By Proposition 5.2, μ_Λ is ergodic, and so from (5.1) and the ergodic theorem (see Walters [Wa]) it follows

$$\lim_{n\to\infty}\frac{1}{n}\log\mathcal{J}_n(x)=h_{\mu_\Lambda}(F)\ \text{for}\ \mu_\Lambda\text{-almost all}\ x \tag{6.6}$$

since $\mathcal{J}_n(x)=\mathcal{J}(x)\mathcal{J}(Fx)\cdots\mathcal{J}(F^{n-1}x)$.

For any integer $n>0$ and a number $\gamma>0$ define

$$\Gamma_\gamma(n)=\{x\in\Lambda:|\frac{1}{n}\log\mathcal{J}_n(x)-h_{\mu_\Lambda}(F)|>\gamma\}. \tag{6.7}$$

If $x\in\Gamma_{2\gamma}(n)$ and $\rho\leq\tilde{\rho}$

$$n>\frac{1}{\gamma}\log C \tag{6.8}$$

then by (3.14),

$$K_\rho(x,n) \cap \Lambda \subset \Gamma_\gamma(n), \tag{6.9}$$

provided $\tilde{\rho}, C > 0$ are the same as in (3.14).

Since according to Proposition 5.2 μ_Λ has conditional measures on unstable manifolds which are equivalent to the Riemannian volume then one can derive that

$$m(K_\rho(x,n)) \le C_\rho \mu_\Lambda(K_\rho(x,n)) \tag{6.10}$$

for some $C_\rho > 0$ independent of $n = 1,2,\ldots$ and $x \in \Lambda$ (see Bowen and Ruelle [BR], Corollary 4.6). Now let $\mathcal{E}(\rho,n)$ be a maximal (ρ,n)-separated set then by (3.13) and (6.7)-(6.10) it follows that

$$\mu_\Lambda(\Gamma_\gamma(n)) \ge \mu_\Lambda\Big[\bigcup_{x\in\mathcal{E}(\rho,n)\cap\Gamma_{2\gamma}(n)} B_\rho(x,n)\Big] \tag{6.11}$$

$$\ge C_{\rho/2}^{-2} \sum_{x\in\mathcal{E}(\rho,n)\cap\Gamma_{2\gamma}(n)} \mathcal{J}_n^{-1}(x).$$

Let now $\Pi = (V_1,\ldots,V_k)$ be a partition satisfying (6.1) with $\rho_0 = \frac{1}{3} C_1^{-1}\rho \le \frac{1}{3}\rho$ where C_1 is the same as in (6.5). Define $U_\rho^\Pi(\Lambda) = \bigcup_{V_i\in\Pi_\rho(\Lambda)} V_i$. Then

$$\Lambda \subset U_{\rho_0}(\Lambda) \subset U_{\rho_0}^\Pi(\Lambda) \subset U_\rho(\Lambda). \tag{6.12}$$

Take $n(\epsilon)$ as in (4.3) and an integer $N > 0$, then by the Markov property

$$\sigma^\epsilon(i_0,\ldots,i_{Nn(\epsilon)}) \tag{6.13}$$

$$\leq \sup_z P_z^\epsilon\{X_1^\epsilon \in V_{i_1},\ldots,X_{n(\epsilon)}^\epsilon \in V_{i_{n(\epsilon)}}\}$$

$$\sup_z P_z^\epsilon\{X_1^\epsilon \in V_{i_{n(\epsilon)+1}},\ldots,X_{n(\epsilon)}^\epsilon \in V_{i_{2n(\epsilon)}}\} \cdots$$

$$\cdots \sup_z P_z^\epsilon\{X_1^\epsilon \in V_{i_{(N-1)n(\epsilon)+1}},\ldots,X_{n(\epsilon)}^\epsilon \in V_{i_{Nn(\epsilon)}}\}$$

where sup is taken over $z \in \overline{U}_\Lambda$. Therefore

$$H_{N\cdot n(\epsilon)}^\epsilon(\Pi) \tag{6.14}$$

$$= - \sum_{i_0,\ldots,i_{N\cdot n(\epsilon)}} \sigma^\epsilon(i_0,\ldots,i_{N\cdot n(\epsilon)}) \log \sigma^\epsilon(i_0,\ldots,i_{N\cdot n(\epsilon)})$$

$$\geq - \sum_{k=1}^N \Big(\sum_{i_{(k-1)n(\epsilon)+1},\ldots,i_{kn(\epsilon)}}$$

$$\sigma^\epsilon(i_{(k-1)n(\epsilon)+1},\ldots,i_{kn(\epsilon)}) \log \sup_z$$

$$P_z^\epsilon\{X_1^\epsilon \in V_{i_{(k-1)n(\epsilon)+1}},\ldots,X_{n(\epsilon)}^\epsilon \in V_{i_{kn(\epsilon)}}\}\Big)$$

$$= -N \sum_{i_1,\ldots,i_{n(\epsilon)}} \sigma^\epsilon(i_1,\ldots,i_{n(\epsilon)}) \log \sup_z$$

$$P_z^\epsilon\{X_1^\epsilon \in V_{i_1},\ldots,X_{n(\epsilon)}^\epsilon \in V_{i_{n(\epsilon)}}\} \overset{\text{def}}{\equiv} N\tilde{H}_{n(\epsilon)}^\epsilon(\Pi).$$

To proceed any further we shall need the following.

Lemma 6.1. *There exist* $\rho_1, C_2, \alpha_1 > 0$ *such that if* $\rho \leq \rho_1$ *then for any* $x \in \Lambda$ *and* $z \in U_\Lambda$,

$$P_z^\epsilon\{ \max_{1\le j\le n(\epsilon)} \operatorname{dist}(X_j^\epsilon, F^j x) \le 2\rho\} \tag{6.15}$$

$$\le C_2 \epsilon^{-\upsilon^u} g^\epsilon(x,z)(\mathcal{J}_{n(\epsilon)}(x))^{-1},$$

provided ϵ *is small enough, where* $g^\epsilon(x,z)$ $= (\epsilon^{(1-18\upsilon\beta)} + e^{-\alpha_1 \|\eta^u\| \epsilon^{-1}})$ *if* $\operatorname{dist}(x,z) \le \epsilon^{1-2\beta}$ *and* $g^\epsilon(x,z) = \exp(-\epsilon^{-\beta/4})$ *otherwise*, $\eta = \operatorname{Exp}_x^{-1} z$, β *satisfies* (4.36), *and we use notations of Section 2.4. Next,*

$$\int_{U_\Lambda} d\mu^\epsilon(dz) P_z^\epsilon\{ \max_{1\le j\le n(\epsilon)} \operatorname{dist}(X_j^\epsilon, F^j x) \le 2\rho\} \tag{6.16}$$

$$\le C_2 (\mathcal{J}_{n(\epsilon)}(x))^{-1}.$$

Proof. By Proposition 3.8 and the transversality of the stable and unstable directions one can choose a constant $C_3 > 0$ such that

$$\tilde{W}^s_{C_3\rho}(E) \supset U_{2\rho}(F^{n(\epsilon)}x), \ E = W^u_{C_3\rho}(x) \tag{6.17}$$

for any $\rho > 0$ small enough, where $\tilde{W}^s_r(\Gamma)$ was defined in Section 2.4 as $W^s_r(\Gamma)$ in the diffeomorphism case and $\tilde{W}^s_r(\Gamma) = \bigcup_{t:|t|\le r} W^s_r(\Gamma)$ in the flow case. In the same way as in Lemma 1.1 and (4.5),

$$P_z^\epsilon\{ \max_{1\le j\le n(\epsilon)} \operatorname{dist}(X_j^\epsilon, F^j x) \le 2\rho\} \tag{6.18}$$

$$= I_2^\epsilon(2\rho, n(\epsilon), z, x, U_{2\rho}(F^{n(\epsilon)}x))$$

$$\leq I_2^{\epsilon}(2\rho, n(\epsilon), z, x, \tilde{W}^s_{C_3\rho}(E))$$

$$\leq I_4^{\epsilon}(2\rho, \delta(\epsilon), n(\epsilon), z, x, A) + \exp(-\epsilon^{-\beta/3})$$

where $\delta(\epsilon) = \epsilon^{1-\beta}$, $A = \tilde{W}^s_{C_3\rho}(E) \cap W^s_{\epsilon^{1-2\beta}}(\Lambda)$, and

$$I_4^{\epsilon}(r, \delta, n, z, x, \Gamma)$$

$$= P_z^{\epsilon}\{ \max_{1 \leq j \leq n(\epsilon)} \operatorname{dist}(X_j^{\epsilon}, F^j x) \leq 2\rho,$$

$$\operatorname{dist}(FX_i^{\epsilon}, X_{i+1}^{\epsilon}) < \delta \text{ for all } i = 0, \dots, n-1,$$

$$\text{and } X_n^{\epsilon} \in \Gamma\}.$$

In the same way as in the proof of Theorem 4.1 one finds points $v_i \in E$, $i = 1, \dots, k_\epsilon$ satisfying (4.10). Then

$$I_4^{\epsilon}(2\rho, \delta(\epsilon), n(\epsilon), z, x, A) \leq \sum_i I_4^{\epsilon}(2\rho, \delta(\epsilon), n(\epsilon), z, x, A_i) \quad (6.19)$$

where $A_i = \tilde{W}^s_{C_2\rho}(E_i) \cap W^s_{\epsilon^{1-2\beta}}(\Lambda)$ and $E_i = W^u_{\epsilon}(v_i)$. Since in

$$I_4^{\epsilon}(2\rho, \delta(\epsilon), n(\epsilon), z, x, \tilde{W}^s_{C_3\rho}(E_i))$$

we have only paths which are $\delta(\epsilon)$-pseudo-orbits $\omega = (z, y_1, \dots, y_{n(\epsilon)})$ starting at z, ending in A_i, and satisfying $y_\ell \in U_{2\rho}(F^\ell x)$ for $\ell = 1,\dots,n(\epsilon)$ then by Proposition 3.3 (or by Proposition 3.4 in the flow case) one can find a point y^ω satisfying (4.12). Thus if ϵ is small enough then

$$y^\omega \in K_{2\rho+\epsilon^{1-2\beta}}(x, n(\epsilon)) \cap U_{\epsilon^{1-2\beta}}(z).$$

It is easy to see from here that there exists a point v_i such that

$$v_i \in W^u_\epsilon(x) \cap K_{2\rho+\epsilon^{1-2\beta}}(x, n(\epsilon)) \cap U_{\epsilon^{1-2\beta}}(z) \tag{6.20}$$

and $y_\ell \in U_{\epsilon^{1-3\beta}}(F^\ell v_i)$ for $\ell = 1,\dots,n(\epsilon)$. But then using (4.27) we derive

$$I^\epsilon_4(2\rho, \delta(\epsilon), n(\epsilon), z, x, A_i) \le I^\epsilon_2(\epsilon^{1-3\beta}, n(\epsilon), z, v_i, A_i) \tag{6.21}$$

$$\le C_4(\mathcal{J}_{n(\epsilon)}(v_i))^{-1}\epsilon^{-\nu^u} m^u(E_i)(e^{-\alpha_1\|\xi^u\|\epsilon^{-1}} + \epsilon^{(1-18\nu\beta)})$$

for some $\alpha_1, C_4 > 0$ where $\xi = \mathrm{Exp}^{-1}_{v_i} z$. By (3.14) and (6.20) it follows

$$C^{-1} \le \mathcal{J}_{n(\epsilon)}(v_i)(\mathcal{J}_{n(\epsilon)}(x))^{-1} \le C \tag{6.22}$$

provided ρ and ϵ are small enough. Thus (4.10) and (6.18)-(6.22) imply

$$P^\epsilon_z\{\max_{1\le j\le n(\epsilon)} \mathrm{dist}(X^\epsilon_j, F^j x) \le 2\rho\} \tag{6.23}$$

$$\le C_5(\mathcal{J}_{n(\epsilon)}(x))^{-1}\epsilon^{-\nu^u} m^u(E)(e^{-\alpha_1\|\eta^u\|\epsilon^{-1}} + \epsilon^{(1-18\nu\beta)})$$

$$+\exp(-\epsilon^{-\beta/3})$$

where $C_5 > 0$ is independent of x, z, and ϵ. Since

$$(\mathcal{J}_{n(\epsilon)}(x))^{-1} \geq (\sup_{x\in\Lambda}\|DF\|)^{-\nu^u n(\epsilon)}, \tag{6.24}$$

the inequality (6.15) follows for small ϵ from the definition (4.3) of $n(\epsilon)$.

To prove (6.16) set $G_j^\epsilon = W^u_{j\epsilon}(x)\setminus W^u_{(j-1)\epsilon}(x)$, then by (4.37) we conclude that

$$\mu^\epsilon(\tilde{W}^s_\rho(G_j^\epsilon)) \leq C_6 m^u(G_j^\epsilon)+2\exp(-\epsilon^{-\beta/3}) \tag{6.25}$$

$$\leq C_7 j^{\nu^u}\epsilon^{\nu^u}+2\exp(-\epsilon^{-\beta/3})$$

for some $C_6, C_7 > 0$ independent of j and ϵ. Now (6.16) follows from (4.3), (6.15), and (6.25). □

Next, we come back to the proof of Theorem 6.1. By Lemma 1.1 for ϵ small enough with probability at least $1-\exp(-\epsilon^{-\beta/2})$ the random points $z, X_1^\epsilon, \ldots, X_{n(\epsilon)}^\epsilon$ form $\delta(\epsilon)$-pseudo-orbits of F. Thus by Proposition 3.3 (or by Proposition 3.4 in the flow case) one can find a random point $y(\omega)$ such that, with probability at least $1-\exp(-\epsilon^{\beta/2})$,

$$\mathrm{dist}(X_k^\epsilon(\omega), F^k y(\omega)) < \epsilon^{1-2\beta} \text{ for all } k = 0,\ldots,n(\epsilon). \tag{6.26}$$

Therefore, for any $z \in \overline{U}_\Lambda$,

$$P_z^\epsilon\{\inf_{y\in U_\rho(z)}\ \sup_{0\leq k\leq n(\epsilon)} \mathrm{dist}(X_k^\epsilon, F^k y) \geq \epsilon^{1-2\beta}\} \tag{6.27}$$

$$\leq \exp(-\epsilon^{-\beta/2}).$$

It follows from (4.6) that

$$\mu^{\varepsilon}(\overline{U}_{\Lambda} \setminus U_{\Lambda}^{\Pi}(\rho_1)) \le \exp(-\varepsilon^{-\beta/3}). \tag{6.28}$$

Now we have

$$\tilde{H}^{\varepsilon}_{n(\varepsilon)}(\Pi) \ge \mathscr{H}^{\varepsilon}(\mathscr{T}^{\gamma}_{n(\varepsilon)}) \tag{6.29}$$

$$\overset{\text{def}}{\equiv} - \sum_{(i_1,\ldots,i_{n(\varepsilon)})\in\mathscr{T}^{\gamma}_{n(\varepsilon)}} \sigma^{\varepsilon}(i_1,\ldots,i_{n(\varepsilon)}) \log \sup_z$$

$$P^{\varepsilon}_z\{X^{\varepsilon}_1 \in V_{i_1},\ldots,V^{\varepsilon}_{n(\varepsilon)} \in V_{i_{n(\varepsilon)}}\},$$

where $\mathscr{T}^{\gamma}_{n(\varepsilon)}$ is the collection of $n(\varepsilon)$-sequences $i_1,\ldots,i_{n(\varepsilon)}$ such that if $(i_1,\ldots,i_{n(\varepsilon)}) \in \mathscr{T}^{\gamma}_{n(\varepsilon)}$ then

(a) $V_{i_j} \cap U_{\rho_0}(\Lambda) \ne 0$, i.e., $V_{i_j} \in \Pi_{\rho_0}(\Lambda)$ for all $j = 1,\ldots,n(\varepsilon)$;

(b) there exists $x \in \mathscr{E}(\rho,n(\varepsilon))$ such that $V_{i_j} \subset U_{2\rho}(F^j x)$ for all $j = 1,\ldots,n(\varepsilon)$, where $\mathscr{E}(\rho,n(\varepsilon))$ is a fixed maximal $(\rho,n(\varepsilon))$-separated set;

(c) the point x from the item (b) does not belong to $\Gamma_{2\gamma}(n(\varepsilon))$ defined by (6.7).

From (6.7) and (6.15) together with the definition (6.29) and (a), (b), (c) above it follows that for some $C_8 > 0$ independent of ε,

$$\frac{1}{n(\epsilon)} \mathscr{H}^{\epsilon}(\mathscr{T}^{\gamma}_{n(\epsilon)}) \geq (h_{\mu_\Lambda}(F)-2\gamma \tag{6.30}$$

$$+(\nu^u \log \epsilon - C_8)(n(\epsilon))^{-1}) \sum_{(i_1,\ldots,i_{n(\epsilon)})\in\mathscr{T}^{\gamma}_{n(\epsilon)}} \sigma^{\epsilon}(i_1,\ldots,i_{n(\epsilon)}).$$

Next by (6.1) with $\rho_0 = \frac{1}{3} C_1^{-1}\rho$, by (6.5), (6.11), (6.16), (6.27), and (6.28) one obtains for ϵ small enough that

$$\sum_{(i_1,\ldots,i_{n(\epsilon)})\notin\mathscr{T}^{\gamma}_{n(\epsilon)}} \sigma^{\epsilon}(i_1,\ldots,i_{n(\epsilon)}) \tag{6.31}$$

$$\leq n(\epsilon)\mu^{\epsilon}(\overline{U}_{\Lambda}\setminus U^{\Pi}_{\Lambda}(\rho_0))$$

$$+\sup_{z\in U_\Lambda} P^{\epsilon}_z\{ \inf_{y\in U_\rho(z)} \sup_{0\leq k\leq n(\epsilon)} \text{dist}(X^{\epsilon}_k, F^k x) \geq \rho_0\}$$

$$+ \sum_{x\in\mathscr{E}(\rho,n(\epsilon))\cap\Gamma_{2\gamma}(n(\epsilon))} \int_{U_\Lambda} d\mu^{\epsilon}(y)$$

$$P^{\epsilon}_y\{ \max_{1\leq j\leq n(\epsilon)} \text{dist}(X^{\epsilon}_j, F^j x) \leq 2\rho\}$$

$$\leq (n(\epsilon)+1)\exp(-\epsilon^{-\beta/3}) + C_8\mu_\Lambda(\Gamma_\gamma(n(\epsilon)))$$

for some $C_9 > 0$ independent of ϵ.

By (6.6),

$$\lim_{\epsilon\to 0} \mu_\Lambda(\Gamma_\gamma(n(\epsilon))) = 0. \tag{6.32}$$

Since $\sum_{i_1,\ldots,i_{n(\epsilon)}} \sigma^{\epsilon}(i_1,\ldots,i_{n(\epsilon)}) = 1$ then (6.30)-(6.32) yield

$$\liminf_{\epsilon\to0} \frac{1}{n(\epsilon)} \mathcal{H}^{\epsilon}(\mathcal{T}^{\gamma}_{n(\epsilon)}) \geq h_{\mu_\Lambda}(F)-2\gamma. \tag{6.33}$$

But γ can be taken arbitrarily small, and so (6.29) implies

$$\liminf_{\epsilon\to0} \frac{1}{n(\epsilon)} \tilde{H}^{\epsilon}_{n(\epsilon)}(\Pi) \geq h_{\mu_\Lambda}(F). \tag{6.34}$$

On the other hand, from the general theory reviewed in Section 1.2 it follows that

$$h^{\epsilon}(\theta,\zeta^{\epsilon}) = \lim_{n\to\infty} \frac{1}{n} H^{\epsilon}_{n}(\Pi) = \inf_{n\geq0} \frac{1}{n} H^{\epsilon}_{n}(\Pi) \tag{6.35}$$

(see (I.2.2) and the definition of $h^{\epsilon}(\theta,\zeta^{\epsilon})$ in Section 1.2), and so by (6.14)

$$h^{\epsilon}(\theta,\zeta^{\epsilon}) = \lim_{n\to\infty} \frac{1}{n} H^{\epsilon}_{n}(\Pi) \geq \frac{1}{n(\epsilon)} \tilde{H}^{\epsilon}_{n(\epsilon)}(\Pi). \tag{6.36}$$

Hence by (6.34),

$$\liminf_{\epsilon\to0} \lim_{n\to\infty} \frac{1}{n} H^{\epsilon}_{n}(\Pi) \geq h_{\mu_\Lambda}(F) \tag{6.37}$$

which together with (6.4) gives (6.3) proving Theorem 6.1. □

Again, disregarding all arguments concerning stable subbundles and stable submanifolds, and using Propositions 3.1 and 3.7, Theorem 4.2 and Theorem 5.4 we obtain in the same way as above the following result.

Theorem 6.2. *Let* F *be a* C^2*-expanding map of a compact Riemannian manifold* M*. Suppose that random perturbations* X^{ϵ}_{n} *satisfy Assumptions 1.1 and 1.2, and let* μ^{ϵ} *denote an invariant measure of* X^{ϵ}_{n} *which according to Theorem 5.4 converges weakly to an F-invariant measure* μ

which is equivalent to the Riemannian volume m. *Then there exists* $\rho_0 > 0$ *such that for any partition* $\Pi = (V_1, \dots, V_k)$ *of* M *into Borel sets satisfying*

$$m(\partial\Pi) = 0 \quad and \quad \max_i \operatorname{diam} V_i \leq \rho_0 \tag{6.38}$$

one has

$$\lim_{\epsilon\to 0} h^{\epsilon}(\theta, \zeta^{\epsilon}) = h_{\mu}(F). \tag{6.39}$$

Remark 6.1. In his talk at Warwick's Symposium on Dynamical Systems in Summer 1986 D. Ornstein announced the following result which he proved together with B. Weiss [OW]. Suppose that both a transformation F considered with its invariant measure μ and random perturbations X_n^{ϵ} considered with invariant measures μ^{ϵ} are Bernoulli systems. Let $\mu^{\epsilon} \xrightarrow{w} \mu$ and $h^{\epsilon}(\theta, \zeta^{\epsilon}) \to h_{\mu}(F)$ as $\epsilon \to 0$ for partitions $\Pi = (V_1, \dots, V_k)$ with $\mu(\partial\Pi) = 0$ and $\max_i \operatorname{diam} V_i \leq \rho_0$ where $\rho_0 > 0$ is small enough. Then for any $\delta > 0$ there is $\epsilon(\delta) > 0$ such that if $\epsilon < \epsilon(\delta)$ the stationary process X_n^{ϵ} taken with an invariant mesure μ^{ϵ} and considered as a dynamical system in the space of paths is isomorphic to a δ-reliable viewer which is defined as follows. The possible states of the viewer form a measure space (V, η), $\eta \in \mathcal{P}(V)$. A Bernoulli measure-preserving transformation $g: V \to V$ governs how the state of the viewer changes in time. If we look at $x \in M$ through the viewer and if the state of the viewer is $v \in V$, then we shall see the point $\Phi(v, x) \in M$ where Φ is a measurable function on $V \times M$. The reliability of the viewer is the expected distance between x and $\Phi(v, x)$ using the product measure $\eta \times \mu$ on $V \times M$. The image under Φ of $\eta \times \mu$ is a measure $\tilde{\mu}$ on M describing what we actually see through the viewer. Next Ornstein and Weiss derive from the general theory of Bernoulli partitions that the stationary process $(X_t^\epsilon, \mu^\epsilon)$ is isomorphic to the dynamical system $(g \times F, V \times M, \tilde{\mu})$ where $\int dist(x, \Phi(v,x)) d\eta(v) d\mu(x) \leq \delta$. Perturbations caused by a δ-reliable viewer they also call δ-harmless since they are not accumulated with time unlike random perturbations X_n^ϵ which are cumulative, i.e., if $X_0^\epsilon = x$ then X_n^ϵ may be far away from $F^n x$ even with probability close to one. They call such random perturbations serious.

Remark 6.2. There are other parameters of dynamical systems corresponding to hyperbolic attractors such as Lyapunov exponents and the dimension with respect to the Sinai-Bowen-Ruelle measure which would be interesting to obtain via random perturbations.

2.7. Stability of the topological pressure.

In this section we shall see that the escape rate of random perturbations from a neighborhood of a hyperbolic set converges as $\epsilon \to 0$ to the topological pressure defined by (5.14). This complements the picture for a hyperbolic set which is not an attractor and it will be important in Section 3.2 for applications to partial differential equations.

Throughout this section Λ will be a hyperbolic set having a local product structure, and so Λ is locally maximal in some open neighborhood $U \supset \Lambda$ (see Definition 3.2 and Remark 3.3). We shall consider both the diffeomorphism and the flow case.

Assumption 7.1. In addition to (1.7) suppose that

$$(1-\epsilon^\alpha)\epsilon^{-\nu} r_x(\epsilon^{-1} \mathrm{Exp}_x^{-1} y) \leq q_x^\epsilon(y) \qquad (7.1)$$

provided $dist(x,y) \leq \epsilon^{1-\alpha}$, and

$$\inf_{x,y\in U} q_x^\epsilon(y) \equiv q^\epsilon > 0 \quad \text{if} \quad \epsilon > 0 \quad \text{is small enough.} \tag{7.2}$$

In the flow case we suppose that $r_y(\xi) = r_y^1(\xi)$ with $r_y^t(\xi)$ given by (1.13).

Remark that (7.1) and (7.2) imply

$$r_x(\xi) > 0 \quad \text{for all} \quad x \in U \quad \text{and} \quad \xi \in T_xM \tag{7.3}$$

which will enable us to apply the lower bound (2.21) in the diffeomorphism case. For the flow case we shall use the lower bound (2.51).

Theorem 7.1. *Under Assumptions 1.1 and 7.1 for any $\rho > 0$ small enough and $x \in U_\rho(\Lambda)$, where Λ is a locally maximal hyperbolic set for a C^2-diffeomorphism F or a C^2-flow F^t (and then $F = F^1$), one has*

$$\lim_{\epsilon\to 0} \limsup_{n\to\infty} \frac{1}{n}\log P_x^\epsilon\{X_k^\epsilon \in U_\rho(\Lambda) \text{ for all } \tag{7.4}$$

$$k = 0,1,\dots,n\} = \lim_{\epsilon\to 0} \liminf_{n\to\infty} \frac{1}{n}\log P_x^\epsilon\{X_k^\epsilon \in U_\rho(\Lambda)$$

$$\text{for all } k = 0,\dots,n\}$$

$$= \lim_{n\to\infty} \frac{1}{n}\log m\{y: F^k y \in U_\rho(\Lambda) \text{ for all } k = 0,\dots,n\}$$

$$= P(\Lambda)$$

where X_n^ϵ are random perturbations of F and m is the Riemannian volume. In the flow case F^t one has also for continuous time random perturbations X_t^ϵ of the diffusion type described in Example 1.4 that

$$\lim_{\epsilon\to 0} \limsup_{T\to\infty} \frac{1}{T}\log P_x^{\epsilon}\{X_t^{\epsilon} \in U_\rho(\Lambda) \text{ for all } t \in [0,T]\} \tag{7.5}$$

$$= \lim_{\epsilon\to 0} \liminf_{T\to\infty} \frac{1}{T}\log P_x^{\epsilon}\{X_t^{\epsilon} \in U_\rho(\Lambda) \text{ for all } t \in [0,T]\}$$

$$= \lim_{T\to\infty} \frac{1}{T}\log m\{y: F^t y \in U_\rho(\Lambda) \text{ for all } t \in [0,T]\} = P(\Lambda).$$

Remark 7.1. We shall see in Section 3.1 that if $U_\rho(\Lambda)$ is replaced by a domain with smooth (piecewise smooth) boundary then $\liminf_{T\to\infty}$ and $\limsup_{T\to\infty}$ can be replaced by $\lim_{T\to\infty}$.

Proof of Theorem 7.1. For any integer $n > 0$ and $x \in U_\rho(\Lambda)$ denote $Z_\rho^{\epsilon}(n,x) = P_x^{\epsilon}\{X_k^{\epsilon} \in U_\rho(\Lambda)$ for all $k = 0,\dots,n\}$. Then $Z_\rho^{\epsilon}(n,x)$ decreases in T and by the Markov property of X_k^{ϵ} one obtains

$$Z_\rho^{\epsilon}(n,x) \le Z_\rho^{\epsilon}(N(\epsilon)n(\epsilon),\rho) \tag{7.6}$$

$$= E_x^{\epsilon}\chi_{X_k^{\epsilon}\in U_\rho(\Lambda) \text{ for all } k=0,\dots,n(\epsilon)}$$

$$\times E^{\epsilon}_{X^{\epsilon}_{n(\epsilon)}}\chi_{X_k^{\epsilon}\in U_\rho(\Lambda) \text{ for all } k=0,\dots,n(\epsilon)}$$

$$\times \cdots \times E^{\epsilon}_{X^{\epsilon}_{(N(\epsilon)-1)n(\epsilon)}}\chi_{X_k^{\epsilon}\in U_\rho(\Lambda) \text{ for all } k=0,\dots,n(\epsilon)}$$

$$\le \Big(\sup_{z\in U_\rho(\Lambda)} Z_\rho^{\epsilon}(n(\epsilon),z)\Big)^{N(\epsilon)}$$

where $n(\epsilon)$ = integral part of $(\log \epsilon)^2$ and $N(\epsilon)$ = integral part of $(n/n(\epsilon))$.

By Lemma 1.1 for ϵ small enough,

$$Z_\rho^\epsilon(n(\epsilon),z) \le I_5^\epsilon(\delta(\epsilon),\rho,n(\epsilon),z) + \exp(-\epsilon^{-\beta/2}) \qquad (7.7)$$

where

$$I_5^\epsilon(\delta,\rho,n,z) = P_z^\epsilon\{X_k^\epsilon \in U_\rho(\Lambda) \text{ for all } k = 0,\dots,n \text{ and}$$

$$\text{dist}(FX_\ell^\epsilon, X_{\ell+1}^\epsilon) < \delta \text{ for all } \ell = 0,\dots,n-1\}.$$

Again, paths $X_0^\epsilon, X_1^\epsilon, \dots, X_n^\epsilon$ appearing in $I_5^\epsilon(\delta(\epsilon),\rho,n(\epsilon),z)$ are $\delta(\epsilon)$-pseudo-orbits $\omega = (z, y_1, \dots, y_{n(\epsilon)})$ staying in $U_\rho(\Lambda)$, and so by Proposition 3.3 (or by Proposition 3.4 in the flow case) if ρ is small enough one can find y^ω satisfying $\text{dist}(F^k y^\omega, y_k) \le \epsilon^{1-2\beta}$ for all $k = 0,\dots,n(\epsilon)$, where $y_0 = z$. Hence $\text{dist}(F^k y^\omega, \Lambda) \le \rho + \epsilon^{1-2\beta}$, $k = 0,\dots,n(\epsilon)$, and so by Proposition 3.5 there exist $\tilde{y}^\omega \in \Lambda$ and a constant $C_1 > 1$ independent of ω such that $y^\omega \in K_{(C_1-1)\rho}(\tilde{y}^\omega, n(\epsilon))$ provided ϵ is small enough.

Let now $\mathcal{E}(\rho,n(\epsilon))$ be a maximal $(\rho,n(\epsilon))$-separated set. Then there exists $\tilde{\tilde{y}} \in \mathcal{E}(\rho,n(\epsilon))$ with $d_{n(\epsilon)}(\tilde{y}^\omega, \tilde{\tilde{y}}^\omega) \le \rho$, and so

$$\text{dist}(F^k \tilde{\tilde{y}}^\omega, y_k) \le C_1\rho \quad \text{for all } k = 0,\dots,n(\epsilon).$$

This leads to the conclusion that

$$I_5^\epsilon(\delta(\epsilon),\rho,n(\epsilon),z) \qquad (7.8)$$

$$\le \sum_{v \in \mathcal{E}(\rho,n(\epsilon))} P_z^\epsilon\{\max_{1\le j\le n(\epsilon)} \text{dist}(X_j^\epsilon, F^j v) \le C_1\rho\}.$$

In the same way as in (6.23) one can see that

$$P_z^\epsilon\{\max_{1\le j\le n(\epsilon)} \operatorname{dist}(X_j^\epsilon, F^j v) \le C_1\rho\} \tag{7.9}$$

$$\le C_2\epsilon^{-\nu^u}(\mathcal{J}_{n(\epsilon)}(v))^{-1} + \exp(-\epsilon^{-\beta/3})$$

$$\le 2C_2\epsilon^{-\nu^u}(\mathcal{J}_{n(\epsilon)}(v))^{-1}$$

for some $C_2 > 0$ independent of ϵ and v. The last inequality in (7.9) follows from (6.24).

Denote $U_\rho(\Lambda,\ell) = \{y : F^i y \in U_\rho(\Lambda)$ for all $i = 0,\dots,\ell\}$ and $K_\rho(\Lambda,\ell) = \bigcup_{y\in\Lambda} K_\rho(y,\ell)$. By Proposition 3.5,

$$K_\rho(\Lambda,\ell) \subset U_\rho(\Lambda,\ell) \subset K_{\tilde{C}\rho}(\Lambda,\ell). \tag{7.10}$$

If $\mathcal{E}(\rho,\ell)$ is a maximal (ρ,ℓ)-separated set then, clearly,

$$\bigcup_{v\in\mathcal{E}(\rho,\ell)} K_{\rho/2}(v,\ell) \subset K_\rho(\Lambda,\ell) \subset \bigcup_{v\in\mathcal{E}(\rho,\ell)} K_{2\rho}(v,\ell) \tag{7.11}$$

and $\{K_{\rho/2}(v,\ell)\}$ are disjoint for different points v from $\mathcal{E}(\rho,\ell)$. Thus it follows that

$$\sum_{v\in\mathcal{E}(\rho,\ell)} m(K_{\rho/2}(v,\ell)) \le m(K_\rho(\Lambda,\ell)) \le m(U_\rho(\Lambda,\ell)) \tag{7.12}$$

$$\le m(K_{\tilde{C}\rho}(\Lambda,\ell)) \le \sum_{v\in\mathcal{E}(\rho,\ell)} m(K_{(\tilde{C}+1)\rho}(v,\ell)).$$

Employing (3.13), Lemma 2.7, Proposition 4.8 from Bowen [Bow], and Proposition 4.4 from Bowen and Ruelle [BR] one obtains

$$P(\Lambda) = \lim_{\ell\to\infty} \frac{1}{\ell}\log m(K_\rho(\Lambda,\ell)) \tag{7.13}$$

$$= \lim_{\ell\to\infty} \frac{1}{\ell}\log m(U_\rho(\Lambda,\ell))$$

$$= \lim_{\ell\to\infty} \frac{1}{\ell}\log \sum_{v\in\mathcal{E}(\rho,\ell)} (\mathcal{J}_\ell(v))^{-1}.$$

Now by (7.6)-(7.9) together with (7.13) we derive

$$\limsup_{\epsilon\to 0}\ \limsup_{n\to\infty} \frac{1}{n}\log Z^\epsilon_\rho(n,x) \tag{7.14}$$

$$\leq \limsup_{\epsilon\to 0} \frac{1}{n(\epsilon)} \log\Big(\sup_{z\in U_\rho(\Lambda)} Z^\epsilon_\rho(n(\epsilon),z)\Big)$$

$$\leq \limsup_{\epsilon\to 0} \frac{1}{n(\epsilon)} \log\Big[\exp(-\epsilon^{-\beta/2})$$

$$+ 2C_2\epsilon^{-\nu^u} \sum_{v\in\mathcal{E}(\rho,n(\epsilon))} (\mathcal{J}_{n(\epsilon)}(v))^{-1}\Big]$$

$$= P(\Lambda).$$

Next we are going to obtain lower estimates. Let $\mathcal{E}(\rho,n)$ be a maximal (ρ,n)-separated set. Then for any $y,z \in \mathcal{E}(\rho,n)$, $y \neq z$ the events $\{\mathrm{dist}(X^\epsilon_k, F^k y) \leq \rho/2$ for all $k = 0,\dots,n\}$ and $\{\mathrm{dist}(X^\epsilon_k, F^k z) \leq \rho/2$ for all $k = 0,\dots,n\}$ are inconsistent. Thus by (7.2),

$$Z^\epsilon_\rho(n,x) \geq Z^\epsilon_\rho(n+1,x) = \int_{U_\rho(\Lambda)} q^\epsilon_{Fx}(y) Z^\epsilon_\rho(n,y)dm(y) \tag{7.15}$$

$$\geq q^\epsilon \int_{U_\rho(\Lambda)} \sum_{v\in\mathcal{E}(\rho,n)} P^\epsilon_y\{\max_{0\leq k\leq n} \mathrm{dist}(X^\epsilon_k, F^k v) \leq \rho/2\}dm(y)$$

$$\geq q^\epsilon \sum_{v\in\mathcal{E}(\rho,n)} \int_{G_\epsilon(v)} P^\epsilon_y\{\max_{0\leq k\leq n} \mathrm{dist}(X^\epsilon_k, F^k v) \leq \rho/2\}dm(y)$$

provided $U \supset U_\rho(\Lambda) \cap FU_\rho(\Lambda)$, where ϵ is so small that

$$G_\epsilon(v) \overset{\text{def}}{=} \mathrm{Exp}_v(\tilde{H}_v^s(\delta(\epsilon))+H_v^u(\epsilon)) \subset U_\rho(\Lambda)$$

where $\tilde{H}_v^s(\delta) = H_v^s(\delta)$ in the diffeomorphism case, $\tilde{H}_v^s(\delta) = H_v^s(\delta)+H_v^0(\delta)$ in the flow case, and $H_v^s(\delta)$, $H_v^0(\delta)$, $H_v^u(\delta)$ are the intersections with the corresponding subbundles of the ball of radius δ centered at the origin of T_vM.

Employing the Markov property one can see for $y \in G_\epsilon(v)$ and ϵ small enough that

$$P_y^\epsilon\{\max_{0\le k\le n} \mathrm{dist}(X_k^\epsilon, F^k v) \le \rho/2\} \tag{7.16}$$

$$\ge \prod_{\ell=0}^{N(\epsilon)} \inf_{z_\ell\in G_\epsilon(v_\ell)} I_2^\epsilon(\rho/2, n(\epsilon), z_\ell, v_\ell, G_\epsilon(F^{n(\epsilon)}v_\ell))$$

where $v_\ell = F^{n(\epsilon)\ell}v$ and the expression $I_2^\epsilon(\cdot,\cdot,\cdot,\cdot,\cdot)$ was introduced in (4.14).

Let $z_\ell \in G_\epsilon(v_\ell)$, $\xi_\ell = \mathrm{Exp}_{v_\ell}^{-1} z_\ell$, and ϵ is small enough. Then making estimates similarly to (4.25)-(4.27) but in the other direction we derive from (4.18) and (7.1) that

$$I_2^\epsilon(\rho/2, n(\epsilon), z_\ell, v_\ell, G_\epsilon(F^{n(\epsilon)}v_\ell)) \tag{7.17}$$

$$\geq (1-\epsilon^\alpha)^{n(\epsilon)} \int_{U_{\epsilon^{1-\beta}}(Fv_\ell)}$$

$$\cdots \int_{U_{\epsilon^{1-\beta}}(F^{n(\epsilon)-1}v_\ell)} \int_{U_{\epsilon^{1-\beta}}(F^{n(\epsilon)}v_\ell)\cap G_\epsilon(F^{n(\epsilon)}v_\ell)}$$

$$\times\ \epsilon^{-\upsilon} r_{Fz_\ell}(\tfrac{1}{\epsilon}\mathrm{Exp}_{Fz_\ell}^{-1} y_1)\epsilon^{-\upsilon} r_{Fy_1}(\tfrac{1}{\epsilon}\mathrm{Exp}_{Fy_1}^{-1} y_2)$$

$$\times \cdots \times\ \epsilon^{-\upsilon} r_{Fy_{n(\epsilon)-1}}(\tfrac{1}{\epsilon}\mathrm{Exp}_{Fy_{n(\epsilon)-1}} y_{n(\epsilon)})dm(y_1)$$

$$\times \cdots \times\ dm(y_{n(\epsilon)})$$

$$\geq (1-\epsilon^\alpha)^{n(\epsilon)}(1-\epsilon^{1-2\beta})^{n(\epsilon)}$$

$$\times \Big[I_3^\epsilon(\delta(\epsilon), n(\epsilon), \xi_\ell, v_\ell, \tilde{H}^s_{v_{\ell+1}}(\delta(\epsilon)) + H^u_{v_{\ell+1}}(\epsilon))$$

$$- \sum_{1\leq i\leq n(\epsilon)} \epsilon^{(1-14\upsilon\beta)i} \prod_{\substack{1\leq j\leq i\\ k_1<\cdots<k_i}} \sup_{\eta\in T_{F^{k_j}v_\ell}(\delta(\epsilon))}$$

$$I_3^\epsilon(\delta(\epsilon), k_{j+1}-k_j-1, \eta, F^{k_j+1}v_\ell, T_{F^{k_j+1}v_\ell}(\delta(\epsilon)))\Big],$$

where the expression I_3^ϵ is the same as in the formula (4.25).

Now we estimate the firt expression I_3^ϵ in (7.17) from below by means of Proposition 2.2 and Proposition 2.4 in the diffeomorphism case or by means of (2.51) and Proposition 2.6 in the flow case. Other appearances of I_3^ϵ (with the negative sign) we estimate from above by virtue of (2.7). These together with (6.24) and (7.17) imply that

$$\inf_{z_\ell \in G_\epsilon(v_\ell)} I_2^\epsilon(\rho/2, n(\epsilon), z_\ell, v_\ell, G_\epsilon(F^{n(\epsilon)} v_\ell)) \tag{7.18}$$

$$\geq C_3 (\mathcal{J}_{n(\epsilon)}(v_\ell))^{-1}$$

for some $C_3 > 0$, and so by (7.16),

$$P_y^\epsilon\{ \max_{0 \leq k \leq n} \operatorname{dist}(X_k^\epsilon, F^k v) \leq \rho/2\} \tag{7.19}$$

$$\geq C_3^{-(N(\epsilon)+1)} \prod_{\ell=0}^{N(\epsilon)} (\mathcal{J}_{n(\epsilon)}(v_\ell))^{-1}$$

$$= C_3^{-(N(\epsilon)+1)} (\mathcal{J}_{(N(\epsilon)+1)n(\epsilon)}(v))^{-1}.$$

Since $m(G_\epsilon(v)) \geq \epsilon^\nu$ if ϵ is small enough then it follows from (3.13), (6.24), (7.13), (7.15), and (7.19) that

$$\liminf_{n\to\infty} \frac{1}{n} \log Z_\rho^\epsilon(n, x) \tag{7.20}$$

$$\geq \liminf_{n\to\infty} \frac{1}{n} \log \sum_{v \in \mathcal{E}(\rho, n)} (\mathcal{J}_n(v))^{-1} = P(\Lambda)$$

which together with (7.14) proves (7.4).

In order to prove (7.5) it suffices only to specify the lower bound. Remark that from the Markov property it is easy to see for $v \in \Lambda$ that

$$P_z^\epsilon\{\sup_{0\le t\le n(\epsilon)} \mathrm{dist}(X_t^\epsilon, F^t v) \le \rho/2 \text{ and } X_{n(\epsilon)}^\epsilon \in \Gamma\} \tag{7.21}$$

$$\ge I_2^\epsilon(C^{-1}\rho, n(\epsilon), z, v, \Gamma)$$

$$-n(\epsilon)(\sup_{w\in\Lambda, y\in U_{C^{-1}\rho}(w)} P_y^\epsilon\{\sup_{0\le t\le 1} \mathrm{dist}(X_t^\epsilon, F^t w) > \rho/2$$

$$\text{and } X_1^\epsilon \in U_{C^{-1}\rho}(F^1 w)\}),$$

with I_2^ϵ introduced in (4.14). We can choose $C > 0$ so that for any $w \in \Lambda$ and $y \in U_{C^{-1}\rho}(w)$,

$$\sup_{0\le t\le 1} \mathrm{dist}(F^t y, F^t w) \le \rho/4 \tag{7.22}$$

and

$$\inf_{0\le s\le 1} \inf_{y:\mathrm{dist}(y,F^s w)>\rho/2} \mathrm{dist}(F^{1-s} y, F^1 w) > 2C^{-1}\rho. \tag{7.23}$$

The large deviation estimates of Wentzell and Freidlin [WF], Theorem 1.2 (see also Friedman [Fri], chapter 14) yield from (7.22) that for any $w \in \Lambda$ and $y \in U_{C^{-1}\rho}(w)$,

$$P_y^\epsilon\{\sup_{0\le t\le 1} \mathrm{dist}(X_t^\epsilon, F^t w) > \rho/2 \text{ and } X_1^\epsilon \in U_{C^{-1}\rho}(F^1 w)\} \tag{7.24}$$

$$\le C_4 \exp\left[-\frac{\gamma}{\epsilon^2}\right] \text{ for some } C_4, \gamma > 0 \text{ independent}$$

of ϵ, w, and y (though depending on ρ which is small but fixed).

One can arrive at (7.24) using simpler (and weaker) estimates of Aronson [Ar1] saying that if $p^\epsilon(t, x_1, x_2)$ is the transition density of the diffusion process X_t^ϵ described in Example 1.4 then

$$p^\epsilon(t,x_1,x_2) \tag{7.25}$$
$$\leq C_5(\epsilon^2 t)^{-\upsilon/2}\exp(-\sigma\epsilon^{-2}(\mathrm{dist}(F^t x_1,x_2))^2)$$

for some C_5, $\sigma > 0$ independent of ϵ, t, x_1, x_2, provided t is bounded and x_1, x_2 vary in a compact set. Let $\tau = \inf\{t:\mathrm{dist}(X_t^\epsilon, F^t w) > \rho/2\}$ then τ is a stopping (Markov) time and by the strong Markov property

$$P_y^\epsilon\{\sup_{0\leq t\leq 1} \mathrm{dist}(X_t^\epsilon, F^t w) > \rho/2 \text{ and } X_1^\epsilon \in U_{C^{-1}\rho}(F^1 w)\} \tag{7.26}$$
$$= E_y^\epsilon \chi_{\tau\leq 1} P^\epsilon(1-\tau, X_\tau, U_{C^{-1}\rho}(F^1 w))$$
$$\leq m(U_{C^{-1}\rho}(F^1 w))\sup\{p(1-s,x_1,x_2):0 \leq s \leq 1,$$
$$\mathrm{dist}(x_1,F^s w) > \rho/2 \text{ and } x_2 \in U_{C^{-1}\rho}(F^1 w)\}.$$

This together with (7.23) and (7.25) imply (7.24).

Now (6.24), (7.18), (7.21) and (7.24) together with the Markov property of the process X_t^ϵ imply

$$P_y^\epsilon\{\sup_{0\leq t\leq n} \mathrm{dist}(X_t^\epsilon, F^t v) \leq \rho/2\} \tag{7.27}$$
$$\geq \prod_{\ell=0}^{N(\epsilon)} \inf_{z_\ell \in G_\epsilon(v_\ell)} P_{z_\ell}^\epsilon\{\sup_{0\leq t\leq n(\epsilon)} \mathrm{dist}(X_t^\epsilon, F^t v_\ell) \leq \rho/2$$
$$\text{and } X_{n(\epsilon)}^\epsilon \in G_\epsilon(F^{n(\epsilon)} v_\ell)\}$$
$$\geq C_6^{-(N(\epsilon)+1)}(\mathcal{J}_{(N(\epsilon)+1)n(\epsilon)}(v))^{-1}$$

and we derive the desired lower bound in the same way as in (7.15) and (7.20). For the upper bound we do not need any continuous time arguments. From these one easily obtains (7.5). For further details we refer the reader to [Ki6]. □

2.8. Appendix: A proof of (1.12).

In this section we shall outline the proof of (1.12) using the parametrix method similar to Aronson [Ar1]. We shall follow [Ki1].

In view of the semigroup structure of both $r_z^{\epsilon}(t,\xi,\eta)$ and $p^{\epsilon}(t,x,y)$ represented by (2.38) and the Chapman-Kolmogorov equality

$$p^{\epsilon}(t+s,x,y) = \int_M p^{\epsilon}(t,x,z)p^{\epsilon}(t,z,y)dm(z) \tag{8.1}$$

it suffices to prove (1.12) only for $t > 0$ small enough, where according to (1.13) and (2.36),

$$\epsilon^{-\nu} r_z(t,\epsilon^1\xi) = r_z^{\epsilon}(t,0,\xi) \tag{8.2}$$

provided we take the Riemannian metric connected with the corresponding differential operator generating the diffusion X_t^{ϵ} as described in Examples 1.3-1.4.

Thus we may assume that the set $\{F^s x, 0 \le s \le t\}$ belongs to the same coordinate chart in M. Moreover, the same arguments which we used in the proof of (7.24) enable us to conclude that with a sufficient precision the asymptotical behavior of $p^{\epsilon}(t,x,y)$ with $\mathrm{dist}(F^t x,y) \le \epsilon^{1-\alpha}$ will not depend on the behavior of the process X_t^{ϵ} outside of a small neighborhood of the set $\{F^s x, 0 \le s \le t\}$. Thus the problem is a local one, and so we can pass from the whole manifold M to a single coordinate chart, and then extend everything to the Euclidean space $\mathbb{R}^{\nu}$, so that $p^{\epsilon}(t,x,y)$ is a fundamental solution in $\mathbb{R}^{\nu}$ of a parabolic equation

$$\frac{\partial p^\epsilon(t,x,y)}{\partial t} = L^\epsilon p^\epsilon(t,x,y) \tag{8.3}$$

where $L^\epsilon = \epsilon^2 L + \langle B,\nabla\rangle$ acts in x, $\langle B(x),\nabla\rangle = \sum_{1\le i\le \nu} B_i(x)\frac{\partial}{\partial x^i}$,

$$L = \frac{1}{2} \sum_{1\le i,j\le \nu} a_{ij}(x)\frac{\partial^2}{\partial x^i \partial x^j} + \sum_{1\le i\le \nu} b_i(x)\frac{\partial}{\partial x^i}\,, \tag{8.4}$$

the matrix $(a_{ij}(x))$ is uniformly positive definite, and all functions $B_i(x)$, $a_{ij}(x)$, $b_i(x)$ $(i,j = 1,\dots,\nu)$ are bounded in $\mathbb{R}^\nu$ together with their first and second derivatives.

The operator L^ϵ generates a diffusion process X_t^ϵ with the transition density $p^\epsilon(t,x,y)$ satisfying (8.3) and $X_t^\epsilon = X_t^\epsilon(x)$ starting at x solves the following stochastic integral equation

$$X_t^\epsilon(x) = x + \int_0^t (B(X_s^\epsilon(x)) + \epsilon^2 b(X_s^\epsilon(x))ds \tag{8.5}$$

$$+ \epsilon \int_0^t \sigma(X_s^\epsilon(x))dw_s$$

where w_s is the ν-dimensional standard Wiener process starting at zero, a matrix function $\sigma(x)$ satisfies $\sigma(x)(\sigma(x))^* = (a_{ij}(x)) \overset{\text{def}}{=} A(x)$, and the star means adjoint.

Let $\mathfrak{D}(s,t,x)$, $s \le t$ be the solution of the matrix integral equation

$$\mathfrak{D}(s,t,x) = I + \int_s^t G(F^u s)\mathfrak{D}(u,s,x)du \quad \text{where} \tag{8.6}$$

$$G(y) = (g_{ij}(y)) = \left[\frac{\partial B_i(y)}{\partial y_j}\right]$$

and I is the identity matrix. Differentiating the equation

$$F^t x = x + \int_0^t B(F^s x)du \tag{8.7}$$

in space variables one concludes that the solution of (8.6) can be expressed by means of the Jacobian matrices, i.e., differentials $DF^u y$ of F^u at y, as

$$\mathcal{D}(s,t,x) = DF^{t-s}_{F^s x} \ . \tag{8.8}$$

Define

$$Y_t(s,x) = \int_s^t \mathcal{D}(u,t,x)\sigma(F^u x)dw_u, \quad t \geq s \tag{8.9}$$

and

$$Z^\epsilon_t(s,x,\xi) = \mathcal{D}(s,t,x)\xi + \epsilon Y_t(s,x), \quad \xi \in R^\nu . \tag{8.10}$$

Evidently, that both $Y_t(s,x)$ and $Z^\epsilon_t(s,x,\xi)$ are nonhomogeneous in time Gaussian processes where $Y_t(s,x)$ has the Gaussian distribution with zero mean and the covariance matrix

$$V(s,t,x) = \int_s^t \mathcal{D}(u,t,x)A(F^u x)(\mathcal{D}(u,t,x))^* du, \tag{8.11}$$

and so $Z^\epsilon_t(s,x,\xi)$ has the Gaussian distribution with the mean $\mathcal{D}(s,t,x)\xi$ and the covariance matrix $\epsilon^2 V(s,t,x)$. Thus we can write the density of the distribution of $Z^\epsilon_t(s,x,\xi)$ at a point $\eta \in R^\nu$ as

$$r_x^\epsilon(s,t,\xi,\eta) = (2\pi\epsilon^2)^{-n/2}(\det V(s,t,x))^{-1/2} \tag{8.12}$$

$$\times \exp\left(-\frac{1}{2\epsilon^2}\langle (V(s,t,x))^{-1}(\eta-\mathfrak{D}(s,t,x)\xi),(\eta-\mathfrak{D}(s,t,x)\xi)\rangle\right)$$

where $\langle\ ,\ \rangle$ denotes the inner product of R^ν.

Using (8.6)-(8.10) one verifies directly that $Z_t^\epsilon(s,x,\xi)$ satisfies the following stochastic integral equation

$$Z_t^\epsilon(s,x,\xi) = \mathfrak{D}(s,t,x)\xi + \int_s^t G(F^u x)Z_u^\epsilon(s,x,\xi)du \tag{8.13}$$

$$+ \epsilon\int_s^t \sigma(F^u x)dw_u .$$

From (8.5) and (8.13) it is easy to see that with probability close to one $|X_t^\epsilon(x)-Z_t^\epsilon(0,x,0)-F^t x|$ is less than $\epsilon^{2-\alpha}$ for any fixed $\alpha > 0$ provided ϵ is small enough. This suggests to compare $p^\epsilon(t,x,y)$ with $r_x^\epsilon(0,t,0,y-F^t x)$. As usual, it is not easy to work with densities by purely probabilistic means, and so we shall employ the partial differential equations parametrix method.

One obtains the relation (1.12) from the following.

Proposition 8.1. *There exists* $C > 0$ *such that if* $|y-F^t x| \le \epsilon^{1-\alpha}$, $0 < \alpha < \frac{1}{3}$, $0 \le t \le 1$ *then*

$$|1-p^\epsilon(t,x,y)(r_x^\epsilon(0,t,0,y-F^t x))^{-1}| \le C\epsilon^{1-3\alpha} \tag{8.14}$$

provided ϵ *is small enough.*

Proof. Consider the operator

$$\mathcal{L}_x^{\epsilon} = \frac{1}{2}\epsilon^2 \sum_{1\leq i,j\leq \nu} a_{ij}(F^s x)\frac{\partial^i}{\partial \xi^i \partial \xi^j} \tag{8.15}$$

$$+ \sum_{1\leq i,j\leq \nu} g_{ij}(F^s x)\xi^i \frac{\partial}{\partial \xi^i}$$

where $G = (g_{ij})$ was defined in (8.6). This operator generates the process $Z_t^{\epsilon}(s,x,\xi)$ in the sense that

$$-\frac{\partial r_x^{\epsilon}(s,t,\xi,\eta)}{\partial s} = \mathcal{L}_x^{\epsilon} r_x^{\epsilon}(s,t,\xi,\eta) \tag{8.16}$$

where $\mathcal{L}_x^{\epsilon}$ acts in the variable ξ (the backward Kolmogorov equation) and this equality can be verified directly.

Denote

$$q^{\epsilon}(s,t,\xi,\eta) = p^{\epsilon}(t-s,F^s x+\xi,F^t x+\eta). \tag{8.17}$$

Then by (8.3) and (8.7),

$$-\frac{\partial q^{\epsilon}(s,t,\xi,\eta)}{\partial s} = \hat{\mathcal{L}}_x^{\epsilon} q^{\epsilon}(s,t,\xi,\eta) \tag{8.18}$$

where the operator

$$\hat{\mathcal{L}}_x^{\epsilon} = \epsilon^2\left[\frac{1}{2} \sum_{1\leq i,j\leq \nu} a_{ij}(F^s x+\xi)\frac{\partial^2}{\partial \xi^i \partial \xi^j} \right. \tag{8.19}$$

$$\left. + \sum_{1\leq i\leq \nu} b_i(F^s x+\xi)\frac{\partial}{\partial \xi^i}\right]$$

$$+ \sum_{1\leq i\leq \nu} (B_i(F^s x+\xi)-B_i(F^s x))\frac{\partial}{\partial \xi^i}$$

acts in the variable ξ. Put

$$\tilde{\mathscr{L}}_x^\epsilon = \hat{\mathscr{L}}_x^\epsilon - \mathscr{L}_x^\epsilon.$$

Then

$$|\tilde{\mathscr{L}}_x^\epsilon r_x^\epsilon(s,t,\xi,\eta)| \tag{8.20}$$

$$= \Big|\epsilon^2\Big[\frac{1}{2}\sum_{1\le i,j\le\nu}(a_{ij}(F^s x+\xi)-a_{ij}(F^s x))\frac{\partial^2 r_x^\epsilon(s,t,\xi,\eta)}{\partial\xi^i\partial\xi^j}$$

$$+\sum_{1\le i\le\nu} b_i(F^s x+\xi)\frac{\partial r_x^\epsilon(s,t,\xi,\eta)}{\partial\xi^i}\Big]$$

$$+\sum_{1\le i\le\nu}\Big[B_i(F^s x+\xi)-B_i(F^s x)-\sum_{1\le j\le\nu} g_{ij}(F^s x)\xi^j\frac{\partial r_x^\epsilon(s,t,\xi,\eta)}{\partial\xi^i}\Big]\Big|$$

$$\le C_1\Big[\epsilon^2|\xi|\max_{i,j}\Big|\frac{\partial^2 r_x^\epsilon(s,t,\xi,\eta)}{\partial\xi^i\partial\xi^j}\Big|+(\epsilon^2+|\xi|^2)\max_i\Big|\frac{\partial r_x^\epsilon(s,t,\xi,\eta)}{\partial\xi^i}\Big|\Big]$$

$$\le C_2 r_x^\epsilon(s,t,\xi,\eta)\Big[\frac{|\xi|}{(t-s)}+\frac{|\xi||\eta-\mathscr{D}(s,t,x)\xi|^2}{\epsilon^2(t-s)^2}$$

$$+\frac{|\eta-\mathscr{D}(s,t,x)\xi|}{(t-s)}\Big(1+\frac{|\xi|^2}{\epsilon^2}\Big)\Big]$$

for some $C_1, C_2 > 0$ independent of ϵ, s, t, ξ, η.

In accordance with the parametrix method, we seek $q^\epsilon(s,t,\xi,\eta)$ in the form

$$q^\epsilon(s,t,\xi,\eta) \tag{8.21}$$

$$= r_x^\epsilon(s,t,\xi,\eta) + \int_s^t\int_{\mathbb{R}^\nu} r_x^\epsilon(s,u,\xi,\zeta)\Phi^\epsilon(u,t,\zeta,\eta)du d\zeta$$

where Φ^ϵ must be determined by the condition that q^ϵ satisfies (8.18). Thus applying the operator $(\frac{\partial}{\partial s}+\hat{\mathscr{L}}_x^\epsilon)$

in the variables s and ξ to both parts of (8.21) we obtain

$$\Phi^{\epsilon}(s,t,\xi,\eta) = (\frac{\partial}{\partial s} + \hat{\mathcal{L}}_x^{\epsilon})r_x^{\epsilon}(s,t,\xi,\eta) \tag{8.22}$$

$$+ \int_s^t \int_{\mathbb{R}^\nu} (\frac{\partial}{\partial s} + \hat{\mathcal{L}}_x^{\epsilon})r_x^{\epsilon}(s,u,\xi,\zeta)\Phi^{\epsilon}(u,t,\zeta,\eta)dud\zeta.$$

In view of (8.16),

$$(\frac{\partial}{\partial s} + \hat{\mathcal{L}}_x^{\epsilon})r_x^{\epsilon}(s,t,\xi,\eta) = \tilde{\mathcal{L}}_x^{\epsilon} r_x^{\epsilon}(s,t,\xi,\eta). \tag{8.23}$$

The solution Φ^{ϵ} of (8.22) can be represented in the form

$$\Phi^{\epsilon}(s,t,\xi,\eta) = \sum_{n=1}^{\infty} (\tilde{\mathcal{L}}_x^{\epsilon} r_x^{\epsilon})_n(s,t,\xi,\eta) \tag{8.24}$$

where $(\tilde{\mathcal{L}}_x^{\epsilon} r_x^{\epsilon})_1 = \tilde{\mathcal{L}}_x^{\epsilon} r_x^{\epsilon}$ and

$$(\tilde{\mathcal{L}}_x^{\epsilon} r_x^{\epsilon})_{n+1}(s,t,\xi,\eta)$$

$$= \int_s^t \int_{\mathbb{R}^\nu} (\tilde{\mathcal{L}}_x^{\epsilon} r_x^{\epsilon}(s,u,\xi,\zeta))(\tilde{\mathcal{L}}_x^{\epsilon} r_x^{\epsilon})_n(u,t,\xi,\zeta)dud\zeta.$$

Indeed, substituting formally (8.24) in (8.22) and taking into account (8.23) we obtain an equality. To justify this one employs the estimate (8.20) and derives the convergence and the differentiability of the series (8.24) in the same way as in Aronson [Ar1]. Finally, using (8.17), (8.21), (8.22), (8.24), and estimating Φ^{ϵ} by means of (8.20) we shall arrive at (8.14). □

Remark 8.1. The relation (8.14) (as well as 1.12) follows also from the precise asymptotics of $p^{\epsilon}(t,x,y)$ obtained in Kifer [Ki3] which has the form

$$p^{\epsilon}(t,x,y) \underset{\epsilon\to 0}{\sim} (2\pi\epsilon^2 t)^{-n/2} K(t,x,y) \exp\left[-\frac{V_t(x,y)}{2\epsilon^2}\right] \quad (8.25)$$

where $K(t,x,y)$ is a bounded positive continuous function and

$$V_t(x,y) = \inf_{\varphi_0=x,\,\varphi_t=y} \int_0^t \|B(\varphi_s)-\dot{\varphi}_s\|^2 ds$$

(which was denoted in (I.58) by $B_t(x,y)$) where the infinum is taken over absolutely continuous curves φ_s, $0 \leq s \leq t$ starting at x and ending at y, $\dot{\varphi}_s = \frac{d\varphi_s}{ds}$, and $\|\cdot\|$ is the Riemannian norm corresponding to the metric form $\sum_{i,j} a^{ij}(x)dx_i dx_j$ where $(a^{ij}(x)) = (a_{ij}(x))^{-1}$ is the inverse matrix to $A(x)$.

Chapter III
Applications to Partial Differential Equations

In this chapter we shall study the asymptotical behavior of eigenvalues of elliptic differential operators generating diffusion perturbations of flows. For some applications to boundary value problems we refer the reader to Kifer [Ki7] and Eizenberg [Ei]. Approaches to other situations can be found in Freidlin and Wentzell [FW].

3.1. Principal eigenvalue and invariant sets.

In this section we shall specify the results of Section 1.3 for the case of diffusion perturbations of flows and connect these with the asymptotical behavior of principal eigenvalues for corresponding generators. We shall follow [Ki5] and [Ki8].

Let $F^t: M \longrightarrow M$ be a C^2 flow on a connected ν-dimensional locally compact Riemannian manifold M and let

$$\frac{dF^tx}{dt} = B\,(F^tx),\ F^0x = x. \tag{1.1}$$

where B(x) is a vector field, and so it can be considered as a first order differential operator. Let L be a differential operator having in local coordinates the form

$$L = \frac{1}{2} \sum_{1\leq i,j\leq \nu} a_{ij}(x) \frac{\partial^2}{\partial x^i \partial x^j} + \sum_{1\leq i\leq \nu} b_i(x) \frac{\partial}{\partial x^i}$$

with C^2 coefficients where the matrix $A(x) = (a_{ij}(x))$ is positive definite for each x. Remark that such operators can be always represented in an invariant form as the sum of the Laplace-Beltrami operator

$$\Delta^{(a)} = \det A(x) \sum_{i,j=1}^{\nu} \frac{\partial}{\partial x^i} \left(\sqrt{\det(a^{ij}(x))}\, a^{ij}(x) \frac{\partial}{\partial x^i}\right)$$

corresponding to the metric form $\sum_{i,j} a^{ij}(x)\, dx_i dx_j$ where $(a^{ij}(x)) = (A(x))^{-1}$, and a vector field considered as a differential operator of the first order.

We shall study diffusion processes X_t^ϵ generated by the operators $L^\epsilon = \epsilon^2 L + B$ in the sense that transition densities $p^\epsilon(t,x,y)$ satisfy Kolmogorov's equation

$$\frac{\partial p^\epsilon(t,x,y)}{\partial t} = L^\epsilon p^\epsilon(t,x,y) \tag{1.2}$$

where L^ϵ acts in the variable x. Such processes X_t^ϵ solve Ito's stochastic differential equations written in local coordinates in the form

$$dX_t^\epsilon = \epsilon\sigma(X_t^\epsilon)dw_t + \epsilon^2 b(X_t^\epsilon)dt + B(X_t^\epsilon)dt \tag{1.3}$$

where $\sigma(x)(\sigma(x))^* = A(x)$ and w_t is the standard ν-dimensional Wiener process (see Friedman [Fri], vol. 1 and Ikeda and Watanabe [IW]). We shall always deal with compact domains in M, and so we shall not need to assume uniform estimates on the differentiable structure of M and the coefficients of the operators L and B.

Let $G \subset M$ be an open set with a smooth or piece-wise smooth boundary ∂G and the compact closure $\bar{G} = G \cup \partial G$. We shall consider the processes X_t^ϵ until the first exit $\tau = \inf\{t: X_t^\epsilon \notin G\}$ from G.

This leads to the diffusion processes X_t^ϵ with the absorption on ∂G whose transition densities $p^\epsilon(t,x,y)$ with respect to the Riemannian volume m satisfy (1.2) in G together with the condition $p^\epsilon(t,x,y)\Big|_{x\epsilon\partial G} = 0$.

Consider the operator P_t^ϵ defined by

$$P_t^\epsilon g(x) = \int_G p^\epsilon(t,x,y)g(y)dm(y), \tag{1.4}$$

acting on the space of continuous functions with zero data on ∂G and with the norm

$$\|g\| = \sup_{x\in G} |g(x)|, \quad \|P_t^\epsilon\| = \sup_{g:\|g\|=1} \|P_t^\epsilon g\|.$$

Since $p^\epsilon(t,x,y)$ is continuous in both variables x and y for any $t > 0$ then P_t^ϵ is a completely continuous operator, and so its spectrum consists of at most countable set of numbers which may accumulate only to zero (see, for instance, Yosida [Yos], Chapter 10). The operators P_t^ϵ form a semigroup $P_{t+s}^\epsilon = P_t^\epsilon P_s^\epsilon$ whose generator in view of (1.2) is the differential operator L^ϵ acting on C^2 functions in G with zero boundary values on ∂G, i.e.,

$$\frac{dP_t^\epsilon g}{dt} = L^\epsilon P_t^\epsilon g \tag{1.5}$$

(see Yosida [Yos], Chapter 9). In view of the semigroup property of operators P_t^ϵ it is easy to see that their eigenvalues have the form $e^{\lambda_i^\epsilon t}$ for some complex numbers λ_i^ϵ, $i = 0,1,\ldots$, and corresponding eigenfunctions g_i^ϵ do not depend on t. Differentiating the equality $P_t^\epsilon g_i^\epsilon = e^{\lambda_i^\epsilon t} g_i^\epsilon$ at t=0 we derive by (1.5) that λ_i^ϵ are eigenvalues of the operator L^ϵ. Using the representation of the generator L^ϵ by means of the resolvent operators (see Yosida [Yos], Chapter 9) we conclude that the spectrum of L^ϵ consists of numbers λ_i^ϵ only.

Denote by λ^ϵ the eigenvalue of L^ϵ with the greatest real part which is called the principal eigenvalue of L^ϵ corresponding to zero Dirichlet data on ∂G. The following result describes the probabilistic sense of λ^ϵ as the speed of absorption of the process X_t^ϵ on the boundary ∂G.

__Theorem 1.1__. *For any* $x \epsilon G$,

$$\lambda^\epsilon = \lim_{t\to\infty} \frac{1}{t} \|P_t^\epsilon\| = \lim_{t\to\infty} \frac{1}{t} P_x\{\tau > t\} \tag{1.6}$$

where τ *is the exit time from* G *and* $P_x\{\cdot\}$ *denotes the probability of an event in brackets for the process* X_t^ϵ *starting at* x.

Proof. It follows from the theory of positive operators (see Krasnoselskii [Kr], p. 259, Theorem 7.10) that the principal eigenvalue λ^ϵ of L^ϵ is simple, real, and negative and the corresponding normilized eigenfunction $r^\epsilon(x)$ is real and positive in G. If $g^\epsilon(t,x) = e^{\lambda^\epsilon t} r^\epsilon(x)$ then

$$\frac{\partial g^\epsilon(t,x)}{\partial t} = e^{\lambda^\epsilon t}\lambda^\epsilon r^\epsilon(x) = e^{\lambda^\epsilon t} L^\epsilon r^\epsilon(x) = L^\epsilon g(t,x) \tag{1.7}$$

and $g^\epsilon(o,x) = r^\epsilon(x)$. Thus $g^\epsilon(t,x)$ can be represented in the form $g^\epsilon(t,x) = P_t^\epsilon r^\epsilon(x)$, and so $e^{\lambda^\epsilon t}$ is the eigenvalue of the operator P_t^ϵ with the positive eigenfunction $r^\epsilon(x)$.

We claim that P_t^ϵ is u_t-positive operator where $u_t(x)=\int_G p^\epsilon(t,x,y)dm(y)$ and recall, $p^\epsilon(t,x,y)$ satisfies (1.2) and the condition $p^\epsilon(t,x,y)|_{x\in\partial G} = 0$. This means by the definition that for any continuous $f \geq 0$, $f\neq 0$, $f|_{\partial G} = 0$ there exists a positive constant $\gamma_f < \infty$ such that

$$\gamma_f^{-1}\, u_t(x) \leq P_t^\epsilon f(x) \leq \gamma_f u_t(x) \text{ for any } x\in G. \qquad (1.8)$$

Indeed, both $P_t^\epsilon f(x)$ and $u_t(x)$ are solutions of the parabolic equation

$$\frac{\partial v(t,x)}{\partial t} = L^\epsilon v(t,x), \quad v|_{\partial G} = 0, \qquad (1.9)$$

they are smooth and positive in G, and by the strong maximum principle (see Protter and Weiberger [PW], Chapter 3, Theorem G) outward normal derivatives are strictly negative along ∂G. It follows that both $(P_t^\epsilon f(x))/u_t(x)$ and $u_t(x)/P_t^\epsilon f(x)$ are bounded for fixed t proving (1.8).

Now we can apply results of Sections 2.2 and 2.3 from Krasnoselskii [Ku] to the operator P_t^ϵ to conclude that P_t^ϵ has the unique positive eigenfunction. This eigenfunction corresponds to the eigenvalue having the maximal absolute value among all eigenvalues of P_t^ϵ. Since by (1.7), $e^{\lambda^\epsilon t}$ is the eigenvalue of P_t^ϵ having this property then $e^{\lambda^\epsilon t}$ must be equal to the spectral radius of the operator P_t^ϵ, which proves the first equality in (1.6).

Next, we shall show that

$$\|P_t^\epsilon\| = \sup_{x\in G} P_x^\epsilon\{\tau>t\}. \qquad (1.10)$$

Since X_t^ϵ denotes here the process with the absorption on ∂G we can write

$$P_x\{\tau>t\} = P_x\{X_t^\epsilon \in G\} = \int_G p^\epsilon(t,x,y)dm(y). \qquad (1.11)$$

Then by (1.4),

$$\sup_{x\in G} P_x^\epsilon\{X_t^\epsilon \in G\} \geq \|P_t^\epsilon\|. \qquad (1.12)$$

Set $f_n(x) = 1$ if dist $(x,\partial G) \geq 2/n$, $x\in G$; $f_n(x) = 0$ if dist $(x,\partial G) \leq \frac{1}{n}$; and $f_n(x) = \frac{1}{2} - \frac{1}{2}\cos(\rho n - 1)\pi$ if dist $(x,\partial G) = \rho$ and $2/n > \rho > \frac{1}{n}$. Put

$$G_n = \{x: \ x\in G \text{ and dist } (x,\partial G) \geq 2/n\}$$

then, as $n \longrightarrow \infty$,

$$\|P_t^\epsilon\| \geq (\sup_{x\in G} \int_G p^\epsilon(t,x,y)f_n(y)dm(y)) \geq \qquad (1.13)$$

$$\geq \sup_{x\in G} \int_G p^\epsilon(t,x,y)dm(y)$$

$$\geq \sup_{x\in G} P_x^\epsilon\{X_t^\epsilon \in G\} - \sup_{x\in G} \int_{G\setminus G_n} p^\epsilon(t,x,y)dm(y)$$

$$\geq \sup_{x\in G} P_x^\epsilon\{X_t^\epsilon \in G\} - m(G\setminus G_n) \sup_{x,y\in G} p^\epsilon(t,x,y)$$

$$\xrightarrow[n\to\infty]{} \sup_{x\in G} P_x^\epsilon \{X_t^\epsilon \in G\}.$$

From (1.11) - (1.13) the assertion (1.10) follows.

Furthermore,

$$P_x^\epsilon\{X_t^\epsilon \in G\} = \int_G p^\epsilon(1,x,y)P_y^\epsilon\{X_{t-1}^\epsilon \in G\}dm(y),$$

and so

$$\inf_{z\in G} p^{\epsilon}(1,x,z) \int_G P_y^{\epsilon}\{X_{t-1}^{\epsilon}\in G\}dm(y) \le P_x^{\epsilon}\{X_t^{\epsilon}\in G\} \tag{1.14}$$

$$\le \sup_{z\in G} p^{\epsilon}(1,x,z) \int_G P_y^{\epsilon}\{X_{t-1}^{\epsilon}\in G\}dm(y).$$

From the first equality in (1.6), which we have already proved, from (1.10), (1.11), and from the right hand side of (1.14) it follows that

$$\lambda^{\epsilon} = \lim_{t\to\infty} \frac{1}{t} \log(\sup_{x\in G} P_x^{\epsilon}\{X_t^{\epsilon}\in G\}) \tag{1.15}$$

$$\le \liminf_{t\to\infty} \frac{1}{t} \log (\int_G P_y^{\epsilon}\{X_{t-1}^{\epsilon}\in G\}dm(y).$$

Similarly, from the left hand side of (1.14) we obtain

$$\lim_{t\to\infty} \frac{1}{t} \log (\sup_{x\in G} P_x^{\epsilon}\{X_t^{\epsilon}\in G\}) \tag{1.16}$$

$$\ge \limsup_{t\to\infty} \frac{1}{t} \log P_z^{\epsilon} \{X_t^{\epsilon}\in G\}$$

$$\ge \limsup_{t\to\infty} \frac{1}{t} \log (\int_G P_y^{\epsilon}\{X_{t-1}^{\epsilon}\in G\}),$$

for any $z\in G$. The relations (1.14) - (1.16) together with (1.10) complete the proof of (1.6). □

Next, we shall improve the result of Section 1.3 for diffusion random perturbations X_t^{ϵ}. Again we shall denote by $\Lambda(\overline{G})$ the maximal F^t-invariant subset of $\overline{G} = G \cup \partial G$ which means that any set $V \subset G \cup \partial G$ satisfying

$$F^t V = V \text{ for all } t\in(-\infty,\infty) \tag{1.17}$$

must be a subset of $\Lambda(\overline{G})$.

Theorem 1.2. *The flow* F^t *has no invariant set in* $\overline{G}$, *i.e.,* $\Lambda(\overline{G}) = \phi$, *if and only if*

$$\lim_{\epsilon\to 0} \lambda^{\epsilon} = -\infty. \tag{1.18}$$

Proof. If $\Lambda(\overline{G}) = \phi$ then by Theorem I.3.1 and Theorem 1.1 we obtain $\limsup_{\epsilon\to 0} \lambda^{\epsilon} = -\infty$ implying (1.18). This follows also from Theorem 7.1 of Chapter 6 in Freidlin and Wentzell [FW] which gives a precise asymptotical behavior of λ^{ϵ} when $\epsilon\to 0$ and $\Lambda(\overline{G}) = \phi$.

Suppose now that (1.18) holds true and prove that $\Lambda(\overline{G}) = \phi$. To do this we assume that $\Lambda(\overline{G}) \neq \phi$ and derive from here that

$$\liminf_{\epsilon\to 0} \lambda^{\epsilon} > -\infty. \tag{1.19}$$

Denote $G_{\rho} = \{x\in G: \ \mathrm{dist}(x,\partial G)>\rho\}$ and, again, $U_{\rho}(z)=\{y: \mathrm{dist}(z,y) < \rho\}$. Let $x\in\Lambda(\overline{G})\subset\overline{G}$ then $F^t x\in G \cup \partial G$ for all t. We can choose numbers ϵ_0, $\delta > 0$ such that if $0<\epsilon\leq\epsilon_0$ there exist points Q_k, $k = 0,1,2,\ldots$ satisfying

$$\mathrm{dist}(Q_k, F^{k\delta}x) = 3\epsilon \text{ and } \min_{0\leq t\leq\delta} \mathrm{dist}\,(F^t Q_k, \partial G) > 2\epsilon. \tag{1.20}$$

The main step in the proof of Theorem 1.2 is the following

Lemma 1.1. *There exists* $q > 0$ *such that for all* $k = 0,1,2,\ldots$ *and* $\epsilon<\epsilon_0$ *one has*

$$\inf_{z\in U_{\epsilon}(Q_k)} P_z^{\epsilon}\{X_{\delta}^{\epsilon}\in U_{\epsilon}(Q_{k+1}) \text{ and } \tau>\delta\} \geq q \tag{1.21}$$

where, again, τ *is the exit time from* G.

Proof. A complication we have here is that x or its iterates may belong to ∂G. Since δ and ϵ are small we can assume that each set $\bigcup_{0\le t\le\delta} U_{6\epsilon}(F^{k\delta+t}x)$ is contained in one coordinate neighborhood which enables us to reduce our problem to the following problem in the Euclidean space $\mathbb{R}^{\nu}$. For some numbers $r_3>r_2>r_1>0$ let $V_k^{(1)}$ and $V_k^{(2)}$ denote open balls in $\mathbb{R}^{\nu}$ of radis r_1 and r_2, respectively, such that

$$U_{r_3}(Q) \supset V_k^{(1)} \supset \bar{V}_k^{(2)} \cup \bar{V}_{k+1}^{(2)} \text{ for all } k = 0,1,2,\cdots. \qquad (1.22)$$

Then we have to show that for some $\tilde{q}>0$ independent of ϵ, k, and the choice of $V_k^{(1)}$ and $V_k^{(2)}$,

$$\inf_{z\in\epsilon V_k^{(2)}+F^{k\delta}x} P_z^{\epsilon}\{X_\delta^\epsilon \in \epsilon V_{k+1}^{(2)}+F^{(k+1)\delta}x \text{ and} \qquad (1.23)$$

$$X_t^\epsilon \in \epsilon V_k^{(1)}+F^{k\delta+t}x \text{ for all } t\in[0,\delta]\} \ge \tilde{q}$$

which will imply (1.21).

Together with the process X_t^ϵ satisfying (1.3) consider also the processes $Y_k^\epsilon(t)=X_t^\epsilon - F^{k\delta+t}x$ and $Z_k^\epsilon(t)$, $k = 0,1,2,\cdots$ where $Z_k^\epsilon(t)$ satisfies

$$dZ_k^\epsilon(t)=\epsilon\sigma(Z_k^\epsilon(t)+F^{k\delta+t}x)dw_t+\epsilon^2 b(Z_k^\epsilon(t)+F^{k\delta+t}x)dt. \qquad (1.24)$$

The processes $Y_k^\epsilon(t)$ and $Z_k^\epsilon(t)$ differ just in drift, and so by the Cameron-Martin-Girsanov theorem (see, Chapter 7 in Friedman [Fri]) we obtain

$$P_z^\epsilon\{X_\delta^\epsilon \in \epsilon V_{k+1}^{(2)} + F^{(k+1)\delta}x \text{ and } X_t^\epsilon \in \epsilon V_k^{(1)} + F^{k\delta+t}x \qquad (1.25)$$

$$\text{for all } t\in[0,\delta]\} = P_{z_k,0}^\epsilon\{Y_k^\epsilon(\delta) \in \epsilon V_{k+1}^{(2)}$$

and $Y_k^\epsilon(t) \in \epsilon V_k^{(1)}$ for all $t\in[0,\delta]\}$

$$= E_{z_k,0}\chi_{Z_k^\epsilon(\delta)\in \epsilon V_{k+1}^{(2)}}\chi_{Z_k^\epsilon(t)\in \epsilon V_k^{(1)}\text{ for all } t\in[0,\delta]}\frac{d\mu_{Y_k^\epsilon}}{d\mu_{Z_k^\epsilon}}(Z_k^\epsilon)$$

$$= E^\epsilon_{\epsilon^{-1}z_k,0}\chi_{\mathcal{Z}_k^\epsilon(\delta)\in V_{k+1}^{(2)}}\chi_{\mathcal{Z}_k^\epsilon(t)\in V_k^{(1)}\text{ for all } t\in[0,\delta]}\frac{d\mu_{Y_k^\epsilon}}{d\mu_{Z_k^\epsilon}}(\epsilon\mathcal{Z}_k^\epsilon)$$

where $z_k = z - F^{k\delta}x$, $\mathcal{Z}_k^\epsilon(t) = \epsilon^{-1}Z_k^\epsilon(t)$, $P^\epsilon_{y,0}$ and $E^\epsilon_{y,0}$ denote the probability and the expectation of the nonhomogeneous processes $Y_k^\epsilon(t)$ and $Z_k^\epsilon(t)$ starting at zero time at y, $\mu_{Y_k^\epsilon}$ and $\mu_{Z_k^\epsilon}$ are probability measures in the path space corresponding to the processes Y_k^ϵ and Z_k^ϵ, and

$$\frac{d\mu_{Y_k^\epsilon}}{d\mu_{Z_k^\epsilon}}(\epsilon\mathcal{Z}_k^\epsilon)=\exp\{\frac{1}{\epsilon}\int_0^\delta\sum_{\ell=1}^{\nu}h_\ell(t,\omega)dw_t^\ell-\frac{1}{2\epsilon^2}\int_0^\delta\sum_{\ell=1}^{\nu}h_\ell^2(t,\omega)dt\} \qquad (1.26)$$

where $w_t = (w_t^1,\cdots, w_t^\nu)$ and $h(t,\omega)=(h_1(t,\omega),\cdots,h_\nu(t,\omega))$ is defined by the formula

$$h(t,\omega)=\sigma^{-1}(\epsilon\mathcal{Z}_k^\epsilon(t)+F^{k\delta+t}x)(B(\epsilon\mathcal{Z}_k^\epsilon(t)+F^{k\delta+t}x)-B(F^{k\delta+t}x)).$$

Let θ_k be the exit time for the process $\mathcal{Z}_k^\epsilon(t)$ from $V_k^{(1)}$ then by the well known property of Ito's stochastic integrals (see Friedman [Fri], Theorem 4.2 of Chapter 4),

$$E_{\mathcal{y}}^\epsilon\, \chi_{\theta_k>\delta}\,(\int_0^\delta\sum_{\ell=1}^{\nu}h_\ell(t,\omega)dw_t^\ell)^2 \qquad (1.27)$$

$$\le E_{\mathcal{y}}^\epsilon(\int_0^{\min(\delta,\theta_k)}\sum_{\ell=1}^{\nu}h_\ell(t,\omega)dw_t^\ell)^2$$

$$= E_{\mathcal{y}}^\epsilon\int_0^{\min(\delta,\theta_k)}\sum_{\ell=1}^{\nu}h_\ell^2(t,\omega)dt.$$

Thus by (1.26) and Chebyshev's inequality we derive that for any $\mathcal{y}\in V_k^{(2)}$ and each $a > 0$,

$$P^{\epsilon}_{\chi,0}\{\mathcal{Z}^{\epsilon}_k(t)\in V^{(1)}_k \text{ for all } t\in[0,\delta] \text{ and } \frac{d\mu_{Y^{\epsilon}_k}}{d\mu_{Z^{\epsilon}_k}}(\epsilon\mathcal{Z}^{\epsilon}_k)<e^{-a}\} \quad (1.28)$$

$$= P^{\epsilon}_{\chi,0}\{\theta_k>\delta \text{ and } (\frac{1}{2\epsilon^2}\int_0^{\delta}\sum_{\ell=1}^{n} h^2_{\ell}(t,\omega)dt - \frac{1}{\epsilon}\int_0^{\delta}\sum_{\ell=1}^{n} h_{\ell}(t,\omega)dw^{\ell}_t)>a\}$$

$$\le \frac{M}{a^2},$$

where $M > 0$ depends just on the upper bounds of the norm $\|\sigma^{-1}(z)\|$ and the derivatives of the vector function $B(z)$, but does not depend on ϵ and a.

Remark, that in view of (1.24) $\mathcal{Z}^{\epsilon}_k(t) = \epsilon^{-1}Z^{\epsilon}_k(t)$ solves a stochastic differential equation with uniformly in ϵ nondegenerated coefficients. In other words, the generator of $\mathcal{Z}^{\epsilon}_k(t)$ is an elliptic operator (with coefficients depending on the time) whose ellipticity constant is independent of ϵ (i.e., at any point the matrix of coefficients in second derivatives has eigenvalues sandwiched between two positive constants independent of ϵ and a point). This enables us to employ uniform estimates from below of fundamental solutions of parabolic equations in a bounded domain (i.e., transition densities of $\mathcal{Z}^{\epsilon}_k(t)$) given in Theorem 8 from Aronson [Ar2] which in view of (see (1.22)) $\inf_{z\notin V^{(1)}_k} \text{dist}\,(z,\bar{V}^{(2)}_k \cup \bar{V}^{(2)}_{k+1}) > 0$ lead to

$$P^{\epsilon}_{\chi,0}\{\mathcal{Z}^{\epsilon}_k(\delta)\in V^{(2)}_{k+1} \text{ and } \mathcal{Z}^{\epsilon}_k(t)\in V^{(1)}_k \text{ for } \quad (1.29)$$

$$\text{all } t\in[0,\delta]\} \ge \rho > 0$$

for some constant $\rho > 0$ independent of k,ϵ, and $\chi\in V^{(2)}_k$. Taking $z \in \epsilon V^{(2)}_k + F^{k\delta}x$ in (1.25) we obtain $\epsilon^{-1}z_k\in V^{(2)}_k$. Thus setting $a = 2M/\rho$ one derives (1.23) from (1.25).

(1.28), and (1.29) with $\tilde{q} = \frac{1}{2} \rho \exp(-\sqrt{2M/\rho})$, and (1.21) follows. □

Now we are able to complete the proof of Theorem 1.2. By (1.21) and the Markov property of the process X_t^ϵ for $z \in \epsilon V_0^{(2)} + x$ one has

$$P_z^\epsilon\{\tau>n\delta\} = E_z^\epsilon \chi_{\tau>\delta} E_{X_\delta^\epsilon}^\epsilon \chi_{\tau>\delta} \chi_{\tau>\delta} \cdots E_{X_{(n-1)\delta}^\epsilon}^\epsilon \chi_{\tau>\delta} \tag{1.30}$$

$$\geq E_z^\epsilon \chi_{\tau>\delta} \chi_{X_\delta^\epsilon \in U_\epsilon(Q_1)} E_{X_\delta^\epsilon}^\epsilon \chi_{\tau>\delta} \chi_{X_\delta^\epsilon \in U_\epsilon(Q_2)}$$

$$\times \cdots \times E_{X_{(n-1)\delta}^\epsilon}^\epsilon \chi_{\tau>\delta} \chi_{X_\delta^\epsilon \in U_\epsilon(Q_n)}$$

$$\geq [\inf_{k\geq 0} \inf_{z \in U_\epsilon(Q_k)} P_z^\epsilon\{X_\delta^\epsilon \in U_\epsilon(Q_{k+1}) \text{ and } \tau>\delta\}]^n \geq q^n.$$

This together with (1.6) yields

$$\lambda^\epsilon \geq \delta^{-1} \log q \tag{1.31}$$

provided ϵ is small enough, which contradicts (1.18), and so the proof of Theorem 1.2 is complete. □

We can improve also the assertion (6) of Theorem I.3.1 in the case of diffusion random perturbations employing $u^\epsilon(x) = E_x^\epsilon \tau$ which, by the way, is the solution of the boundary value problem (see Friedman [Fri], Section 5 of Chapter 6)

$$L^\epsilon u^\epsilon = -1, \quad u^\epsilon|_{\partial G} = 0. \tag{1.32}$$

Theorem 1.3 (a) *If for some* $x \in G$,

$$\limsup_{\epsilon \to 0} E_x^\epsilon \tau = \infty, \tag{1.33}$$

then $\Lambda(\overline{G}) \neq \phi$;

(b) If for any $x \in G$,

$$\liminf_{\epsilon \to 0} E_x^{\epsilon} \tau < \infty, \tag{1.34}$$

then the open domain G *contains no invariant with respect to* F^t *closed subset;*

(c) The item (b) cannot be improved, i.e., the case when for any $x \in G$

$$\limsup_{\epsilon \to 0} E_x^{\epsilon} \tau < \infty \tag{1.35}$$

and $\Lambda(\overline{G}) \neq \phi$ *is possible (i.e., in the closure* $\overline{G}$ *of* G *an invariant subset may exist).*

Proof. The assertion (a) follows immediately from the item (b) of Theorem 1.3.1.

Suppose now that G contains an invariant closed subset Λ. Then $\delta = \inf_{x \in \Lambda} \text{dist}(x, \partial G) > 0$, and so

$$\inf_{-\infty < t < \infty} \text{dist}(F^t x, \partial G) \geq \delta \text{ for any } x \in \Lambda.$$

Employing standart extimates for stochastic integrals it is not difficult to show (see, Theorem 1.2 in Chapter 2 of Freidlin and Wentzell [FW]) that

$$P_x\{\sup_{0 \leq t \leq T} \text{dist}(X_t^{\epsilon}, F^t x) \geq \frac{\delta}{2}\} \longrightarrow 0 \text{ as } \epsilon \longrightarrow 0 \tag{1.36}$$

for any $x \in \overline{G}$ and $T > 0$. Taking $x \in \Lambda$ and an integer $N > 0$ we conclude from (1.36) that

$$P_x^{\epsilon}\{\tau < N\} \longrightarrow 0 \text{ as } \epsilon \longrightarrow 0. \tag{1.37}$$

But

$$E_x^\epsilon \tau \geq \sum_{n=1}^{\infty} (n-1)\, P_x^\epsilon\{(n-1) \leq \tau < n\} \geq N P_x^\epsilon\{\tau \geq N\},$$

and so by (1.37),

$$\liminf_{\epsilon \to 0} E_x^\epsilon\, \tau \geq N$$

for any $N > 0$. Therefore

$$\liminf_{\epsilon \to 0} E_x^\epsilon\, \tau = \infty$$

which contradicts (1.34) and proves the item (b) of Theorem 1.3.

Next, we shall describe an example which satisfies (1.35) and $\Lambda(\overline{G}) \neq \phi$. Let

$$G = \{(x_1, x_2): \quad (x_1 - 1)^2 + x_2^2 < 1\} \subset \mathbb{R}^2$$

and

$$L^\epsilon = \frac{1}{2}\epsilon^2\left(\frac{\partial^2}{\partial x_1^2} + \frac{\partial^2}{\partial x_2^2}\right) - x_2 \frac{\partial}{\partial x_1} + x_1 \frac{\partial}{\partial x_2}. \tag{1.38}$$

In this case $\Lambda(\overline{G})$ is the single point (0,0) and for any $x = (x_1, x_2)$,

$$F^t x = \begin{bmatrix} \cos t & -\sin t \\ \sin t & \cos t \end{bmatrix} \begin{bmatrix} x_1 \\ x_2 \end{bmatrix} \quad ,$$

Thus for each $x \in G$ there exists $t(x) \in [0, \pi]$ such that $F^t x \notin \overline{G}$. Let $p^\epsilon(t,x,y)$ be the transition density of the process X_t^ϵ generated by L^ϵ in the whole plane $\mathbb{R}^2$. Then

$$p^\epsilon(t,x,y) = \frac{1}{2\pi\epsilon^2} \exp\left\{-\frac{1}{2\epsilon^2}|y - F^t x|^2\right\}. \tag{1.39}$$

If $y \in U_\epsilon(F^{t(x)}x)$, then by (1.39),

$$p^\epsilon(t(x),x,y) \geq (2\pi\epsilon^2\sqrt{e})^{-1}. \tag{1.40}$$

The area of $U_\epsilon(F^{t(x)}x)\cap(\mathbb{R}^2\backslash G)$ is greater than $\pi\epsilon^2/2$ and by (1.40) we obtain

$$\begin{aligned} P_x^\epsilon\{\tau\le\pi\} \ge P_x^\epsilon\{\tau\le t(x)\} & \qquad (1.41) \\ \ge P_x^\epsilon\{X_{t(x)}^\epsilon\notin G\} & \\ \ge P_x^\epsilon\{X_{t(x)}^\epsilon\in U_\epsilon(F^{t(x)}x)\cap(\mathbb{R}^2\backslash G)\} & \\ \ge (4\sqrt{e})^{-1}. & \end{aligned}$$

Hence $P_x^\epsilon\{\tau>\pi\} \le 1-(4\sqrt{e})^{-1}$ for all $x\epsilon G$, and so using the Markov property we obtain

$$\begin{aligned} P_x^\epsilon\{\tau>n\pi\} & \qquad (1.42) \\ = E_x^\epsilon\chi_{\tau>\pi}\, E_{X_\pi^\epsilon}^\epsilon \chi_{\tau>\pi} \cdots E_{X_{(n-1)\pi}^\epsilon}^\epsilon \chi_{\tau>\pi} & \\ \le (1-(4\sqrt{e})^{-1})^n. & \end{aligned}$$

Finally,

$E_x^\epsilon\tau \le \pi \sum_{n=1}^{\infty} P_x^\epsilon\{\tau > \pi(n-1)\} \le 4\pi\sqrt{e} < \infty$ giving (1.35) in spite of $\Lambda(\bar{G}) = \{0,0\} \ne \phi$. □

3.2 Localization theorem

In this section we establish a general result which enables one to derive the behavior of the principal eigenvalue λ^ϵ as $\epsilon \longrightarrow 0$ in a domain G from the study of principal eigenvalues corresponding to small neighborhoods of eguivalence classes defined in Section 1.4. We shall discuss also the corresponding asymptotics for different types of equivalence classes. Our exposition will be close to Eizenberg and Kifer [EK].

Let M, F^t, X_t^ϵ, and G be the same as in Section 3.1. A sequence of points $x_0,\cdots x_n \in \bar{G}$ satisfying (I.4.1) will be called again a δ-pseudo-orbit. In the same way as in Section 1.4 we shall write $x \longrightarrow y$, $x,y\in\bar{G}$ if for any $\delta > 0$ there exist a nonnegative $t < 1$ and a δ-pseudo-orbit in $\bar{G}$ starting at $F^t x$ and ending at y. Extending "$\longrightarrow$" to a transitive relation we obtain an order relation "$\succ$". If $x\succ y$ and $y\succ x$ we write $x\sim y$ which is an equivalence relation. Any maximal set of equivalent point in $\bar{G}$ is called an equivalence class.

For any open set $V \subset M$ with a piece-wise smooth boundary ∂V let $\lambda^{\epsilon}(V)$ denotes the principal eigenvalue of the operator L^{ϵ} in V corresponding to zero Dirichlet data on ∂V. Then by Theorem 1.1 for any $x\in V$,

$$\lambda^{\epsilon}(V) = \lim_{t\to\infty} \frac{1}{t} \log P_x^{\epsilon}\{\tau(V)>t\} = \lim_{t\to\infty} \frac{1}{t} \log \Phi^{\epsilon}(t,V) \quad (2.1)$$

where

$$\tau(V) = \inf\{t: \ X_t^{\epsilon}\notin V\} \text{ and } \Phi^{\epsilon}(t,V) = \sup_{x\in U} P_x^{\epsilon}\{\tau(V)>t\}.$$

We shall need

Assumption 2.1 There exists a finite collection of F^t-invariant (in the sense of (1.17)) disjoint equivalence classes $K_1,\cdots K_m \subset G \cup \partial G$ (which are, clearly, closed sets) such that

(i) $\bigcup_i K_i$ contains the limit set of the dynamical system, F^t in $\bar{G}$, i.e., for any $x\in\bar{G}$ all limit points of $F^t x$ as $t \longrightarrow \pm\infty$ which belong to $\bar{G}$ belong also to $\bigcup_i K_i$;

(ii) one can choose open disjoint sets $V_i\subset M$, $i=1,\cdots,m$ with smooth boundaries ∂V_i such that $V_i \supset K_i$, the limit

$$\lambda(K_i) = \lim_{\epsilon\to 0} \lim_{t\to\infty} \frac{1}{t} \log \Phi^{\epsilon}(t,V_i\cap G) \quad (2.2)$$

exists, and for some positive $\beta_o < 1$ and each $\delta > 0$ there is $\epsilon(\delta) > 0$ so that if $\epsilon \leq \epsilon(\delta)$ then one can find a positive

$$t(\epsilon,\delta) \leq e^{-2(1-\beta_o)} \tag{2.3}$$

satisfying

$$\Phi^{\epsilon}(t(\epsilon,\delta),\ V_i \cap G) \leq \exp\ (\lambda(K_i) + \delta)t(\epsilon,\delta). \tag{2.4}$$

Now we can formulate the "localization theorem".

Theorem 2.1 *Under Assumption 2.1*

$$\lim_{\epsilon \to 0} \lambda^{\epsilon}(G) = \max_{1 \leq i \leq m} \lambda(K_i) \tag{2.5}$$

and the numbers $\lambda(K_i)$ defined by (2.3) are determined by the compacts K_i only, i.e., they do not depend on the choice of V_i.

From Theorem II.7.1 and the independence of $\lambda(K_i)$ of V_i we derive

Corollary 2.1 *Let $V \subset M$ be an open set with a piecewise smooth boundary ∂V. Suppose that V contains a locally maximal hyperbolic set Λ and there is no other F^t-invariant set contained in $V \cup \partial V$. Then*

$$\lim_{\epsilon \to 0} \lambda^{\epsilon}(V) = P(\Lambda) \tag{2.6}$$

where $P(\Lambda)$ is the topological pressure defined by (II.5.14). Moreover, since the upper bound in Theorem II.7.1 was established for the time $t = n(\epsilon)$ of order $(\log\epsilon)^2$ then (2.4) with the restriction (2.3) is also satisfied.

Remark 2.1. We shall see how the upper bound (2.4) enters our argument with the restriction (2.3). In all known to the author examples this is satisfied. Clearly, (2.4) holds true if $\lambda(K_i)=0$. It would be intersting to understand whether (2.4) with the restriction (2.3) always follows from (2.2) if $\lambda(K_i)>-\infty$.

Remark 2.2. Theorem 2.1 enables one to reduce the problem to the study of principal eigenvalues for the operator L^{ϵ} restricted to small neighborhoods of compacts K_i. In order to do this by probabilistic means one needs to know decreasing rates for probabilities of staying in small neighborhoods of K_i. When K_i is a hyperbolic set this was accomplished in Theorem II.7.1 where $\lambda(K_i)$ turns out to be the topological pressure $P(K_i)$. An independent study for hyperbolic limit points and circles was performed in Kifer [Ki5]. In all these cases the number $\lambda(K_i)$ can be obtained as

$$\lambda(K_i) = \lim_{T\to\infty} \frac{1}{T} \log \text{volume} \{x: \sup_{0\le t\le T} \text{dist} (F^t x, K_i)\le\rho\} \quad (2.7)$$

provided $\rho > 0$ is small enough. The conjecture is, that (2.7) remains true under more general circumstances. If K_i is an isolated fixed point (not necessarily hyperbolic) namely

$$B(x)= \Pi\, (x-K_i) + O((\text{dist}(x,K_i))^2)$$

locally, with a constant matrix Π being the linear part of $B(x)$ near K_i, then it is easy to see straightforwardly that the right hand side of (2.7) equals $-\sum_i \max(\text{Re}\alpha_i, 0)$ where $\alpha_i, i=1,\cdots,\nu$ are eigenvalues of the matrix Π and Re denotes the real part. The formula (2.7) works only when K_i is strictly inside of G. The case $K_i \subset \partial G$ will be discussed at the end of this section.

Remark 2.3. If K_i is an attractor then (2.7) assigns zero value for $\lambda(K_i)$. In this case one may be interested in the speed of convergence of $\lambda^{\epsilon}(G)$ to zero as $\epsilon \to 0$. These problems are discussed in §7 of Chapter 6 from

Freidlin and Wentzell [FW] (see also Friedman [Fri], Chapter 14) where under certain conditions the convergence of $\epsilon^2 \log |\lambda^\epsilon(G)|$ as $\epsilon \to 0$ to a negative constant (depending on the operator L and the domain V) has been established.

Remark 2.4. If $x_0=K_i$ is an isolated fixed point and there is no other F^t-invariant sets in $\bar{V}_i$, int $V_i \supset x_0$ then the author [Ki7] found the asymptotical behavior of the mean exit time from V_i when starting at x_0 which has the form

$$|\log \epsilon|^{-1} E^\epsilon_{x_0} \tau(V_i) \longrightarrow (\max_i(\mathrm{Re}\alpha_i, 0))^{-1} \text{ as } \epsilon \longrightarrow 0$$

where α_i are eigenvalues of the matrix Π introduced in Remark 2.2. When $\max_i \mathrm{Re}\alpha_i \leq 0$ the above formula is too rough. For an attracting point x_0 Freidlin and Wentzell [FW], Sections 5 of Chapter 6 showed that $\epsilon^2 \log E^\epsilon_x \tau(V)$ converges to a positive constant as $\epsilon \longrightarrow 0$.

Proof of Theorem 2.1. First notice that $\tau(G) \geq \tau(V_i \cap G)$. Hence, $P^\epsilon_x\{\tau(G)>t\} > P^\epsilon_x\{\tau(V_i \cap G)>t\}$ for any $x \in G$, and so by (2.1),

$$\lambda^\epsilon(G) \geq \lambda^\epsilon(V_i \cap G) = \lim_{t\to\infty} \frac{1}{t} \log \Phi^\epsilon(t, V_i \cap G). \tag{2.8}$$

Thus by (2.2),

$$\liminf_{\epsilon \longrightarrow 0} \lambda^\epsilon(G) \geq \max_{1\leq i\leq \nu} \lambda(K_i). \tag{2.9}$$

Therefore it remains only to estimate $\lambda^\epsilon(G)$ from above. By the Markov property of the process X^ϵ_t for any open set V one has

$$\begin{aligned}\Phi^\epsilon(t+s, V) &= \sup_{x\in V} P^\epsilon_x\{\tau(V)>t+s\} \\ &= \sup_{x\in V} E^\epsilon_x \chi_{\tau(V)>t} E^\epsilon_{X^\epsilon_t} \chi_{\tau(V)>t} \\ &\leq \Phi^\epsilon(t, V)\Phi^\epsilon(s, V).\end{aligned} \tag{2.10}$$

If V=G then by (2.1) and the standart subadditivity argument (see, for instance, Walters [Wa], p.88) it follows that

$$\lambda^{\epsilon}(G) = \inf_{t>0} \frac{1}{t} \log \Phi^{\epsilon}(t.G). \tag{2.11}$$

For any open set V with a piece-wise smooth boundary ∂V denote by $P_V^{\epsilon}(t,x,\Gamma)$ transition probabilities of processes X_t^{ϵ} with absorption on ∂V. The corresponding transition densities $p_V^{\epsilon}(t,x,y)$, i.e., the Radon-Nikodim derivatives $\frac{P_V^{\epsilon}(t,x,dy)}{dm(y)}$ with respect to the volume turn out to be fundamental solutions of the equations

$$\frac{\partial p_V^{\epsilon}}{\partial t} = L^{\epsilon} p_V^{\epsilon}, \quad p_V^{\epsilon}(t,x,y)\,|_{\partial V}=0. \tag{2.12}$$

where the operator L^{ϵ} is applied in the variable x. Employing the Chapman-Kolmogorov equality and the estimate (II.7.25) we derive in the same way as in Lemma II.1.1 that for $x \in G$,

$$P_x^{\epsilon}\{\tau(G)>n\} = P_V^{\epsilon}(n,x,G) \le P_x^{\epsilon}\{\text{dist }(F^1X_i^{\epsilon},X_{i+1}^{\epsilon})<\delta \text{ for } \tag{2.13}$$

$$\text{all } i=0,\dots,n-1 \text{ and } X_t^{\epsilon}\in G \text{ for all } t\in[0,n]\}$$
$$+ C_1\epsilon^{-2\nu} n \exp(-\beta_1\delta^2\epsilon^{-2})$$

where C_1, $\beta_1>0$ are independent of ϵ, δ, μ, and n. In the right hand side of (2.13) we have δ-psudo-orbits starting at x and staying in $\bar{G}$. This motivates our next step which is the study of possible behaviors of δ-pseudo-orbits under Assumption 2.1(i).

Let K_i, $i=1,\dots,m$ be compacts introduced in Assumption 2.1. We shall write $K_j \succ K_i$ if there exists a pair of points $x \in K_i$ and $y \in K_j$ such that $y \succ x$. Since K_i and K_j are equivalence classes then $K_j \succ K_i$ means that $y \succ x$ for any $x \in K_i$ and $y \in K_j$. Thus $K_j \succ K_i$ and $K_i \succ K_j$ implies $K_i = K_j$, i.e., $i = j$. We shall need

Lemma 2.1. *For any sufficiently small* $\theta > 0$ *there exists a positive* $\delta(\theta) < \theta$ *such that, if for some* $i_1, i_2 \le m$ *one can find a* $\delta(\theta)$*-pseudo-orbit* $x_0, \dots, x_n \in \bar{G}$ *satisfying*

$$\text{dist}\,(x_0, K_{i_1}) \le \delta(\theta),\ \text{dist}(x_j, K_{i_1}) \ge \theta \text{ and}$$
$$\text{dist}(x_n, K_{i_2}) \le \delta(\theta), \tag{2.14}$$

with $1 < j \le n$, *then* $i_1 \ne i_2$ *and* $K_{i_2} \succ K_{i_1}$.

Proof. Suppose that for any $\delta > 0$ there exists a δ-pseudo-orbit $x_0^{(\delta)}, \dots, x_{n(\delta)}^{(\delta)}$ such that

$$\text{dist}\,(x_0^{(\delta)}, K_{i_1}) \le \delta \text{ and } \text{dist}(x_n^{(\delta)}, K_{i_2}) \le \delta. \tag{2.15}$$

Then one can pick up points $y^{(\delta)} \in K_{i_1}$ and $z^{(\delta)} \in K_{i_2}$ satisfying $\text{dist}\,(y^{(\delta)}, x_0^{(\delta)}) \le \delta$ and $\text{dist}\,(z^{(\delta)}, x_{n(\delta)}^{(\delta)}) \le \delta$. Since K_{i_1} and K_{i_2} are F^t-invariant it follows that $y^{(\delta)}, x_1^{(\delta)}, \dots, x_{n(\delta)-1}^{(\delta)}, z^{(\delta)}$ is a $C_2\delta$-pseudo-orbit where

$$C_2 = \sup_{|t| \le 1} \sup_x \|DF^t\|_x + 2, \tag{2.16}$$
$$\|DF^t\|_x = \sup_{\xi \in T_xM,\, \|\xi\|=1} \|DF^t \xi\|,$$

and the norm are generated by the Riemannian metric. If we assume that such pseudo-orbits can be constructed for any $\delta > 0$ small enough then by the definition $K_{i_2} \succ K_{i_1}$. Hence if $K_{i_2} \succ K_{i_1}$ does not hold true then a δ-pseudo-orbit $x_0^{(\delta)}, \dots, x_{n(\delta)}^{(\delta)}$ satisfying (2.15) may only exist for δ bigger than some $\tilde{\delta} > 0$. In other words, the existence of a δ-pseudo-orbit satisfying (2.15) with $\delta \le \tilde{\delta}$ implies already that $K_{i_2} \succ K_{i_1}$.

Now it remains to discuss the case $i_1 = i_2$. Fix $\theta > 0$. It suffices to show that there exists $\tilde{\tilde{\delta}} > 0$ such that any δ-pseudo-orbit $x_0^{(\delta)}, \cdots, x_{n(\delta)}^{(\delta)}$ satisfying (2.15) with $\delta \le \tilde{\tilde{\delta}}$ and $i_1 = i_2$ contains no points whose distance from K_{i_1} is more than θ. Suppose that, on the contrary, one can find a sequence $\delta_\ell \longrightarrow 0$ as $\ell \to \infty$ and corresponding δ_ℓ-pseudo-orbits $x_0^{(\delta_\ell)}, \cdots, x_{n(\delta_\ell)}^{(\delta_\ell)} \in \bar{G}$ satisfying (2.15) with $\delta = \delta_\ell$, $i_1 = i_2$, and $\text{dist}\,(x_{j(\delta_\ell)}^{(\delta_\ell)}, K_{i_1}) \ge \theta$ for some index $j(\delta_\ell)$. Since the sequence $x_{j(\delta_\ell)}^{(\delta_\ell)}$ stays in a compact set and K_{i_1} is compact, as well, we can choose a subsequence, which we denote again by δ_ℓ, such that $x_0^{(\delta_\ell)} \longrightarrow x$, $x_{n(\delta_\ell)}^{(\delta_\ell)} \longrightarrow y$, and $x_{j(\delta_\ell)}^{(\delta_\ell)} \longrightarrow z$. Then it follows from the definition that $y \succ z \succ x$. But this is impossible since K_{i_1} is the equivalence class, $x, y \in K_{i_1}$, and $\text{dist}\,(z, K_{i_1}) \ge \theta$. This completes the proof of Lemma 2.1. □

For any set $\Gamma \subset M$ we shall use the notations $U_\delta(\Gamma) = \{z \in M: \ \text{dist}\,(z, \Gamma) < \delta\}$ and $U_\delta^G(\Gamma) = U_\delta(\Gamma) \cap \bar{G}$. Choose $\theta_o > 0$ such that

$$U^G_{2C_2\theta_o}(K_i) \subset V_i \cup \partial V_i \text{ for all } i=1,\cdots,m \tag{2.17}$$

$$\text{and put } \delta_o = \delta(\theta_o) C_2^{-1}$$

with $\delta(\theta)$ given by Lemma 2.1 and C_2 defined by (2.16). Since the limit set of the dynamical system F^t restricted to $\bar{G}$ is closed, and by Assumption 2.1(i), it is disjoint with $\bar{G} \setminus \bigcup_i K_i$ then there exists $\delta_1 > 0$ small enough such that the set $\overline{U_{\delta_1}(G)} \setminus \bigcup_{1\le i\le m} U_{\frac{1}{2}\delta_o}(K_i)$ has no common points with this limit set. Thus the number

$$t(x) = \inf\{u\geq 0: F^u x \notin \overline{U_{\delta_1}(G)} \setminus \bigcup_{1\le i\le m} U_{\frac{1}{2}\delta_o}(K_i)\} \tag{2.18}$$

is finite for any $x\in\overline{G}$. Furthermore, it is easy to see that $t(x)$ is upper semicontinuous, i.e., $t(x) \geq \limsup_{y\to x} t(y)$, and so

$$T_o = \sup_{x\in\overline{G}} t(x) < \infty \tag{2.19}$$

Clearly, if $x_0,\cdots,x_n$ is a δ-pseudo-orbit then

$$\max_{o\le k\le n} \operatorname{dist}(x_k, F^k x_0) \leq C_2^{n-1}\delta. \tag{2.20}$$

Notice that if $F^t x \in U_{\frac{1}{2}\delta_o}(K_i)$ for some x, t, and i then by the F^t-invariance of K_i it follows that

$$F^{[t]+1} x \in U_{\frac{1}{2}C_2\delta_o}(K_i)$$

where [t] denotes the integral part of t. This together with (2.18) - (2.20) yield

Lemma 2.2. *Any δ-pseudo-orbit $x_0,\cdots x_n \in\overline{G}$ with $n \geq T_o+1$ and $\delta \leq \frac{1}{2}\delta_o C_2^{-(T_o+1)}$ has at least one point in $\bigcup_{1\le i\le m} U^G_{\delta(\theta_o)}(K_i)$.*

The following result makes the order relation among the compacts K_i more transparent.

Corollary 2.1. *The relation $K_j \succ K_i$ holds true if and only if one can find indices $r_1 = i, r_2,\cdots,r_s=j$ and points $y_1,\cdots,y_{s-1}$ such that for all $k=1,\cdots s$*

$$\operatorname{dist}(F^{-t}y_k, K_{r_k}) + \operatorname{dist}(F^t y_k, K_{r_{k+1}}) \longrightarrow 0 \text{ as } t \longrightarrow \infty. \tag{2.21}$$

Proof. It is evident that (2.21) yields $K_j > K_i$, and so we shall deal only with the inverse implication. Suppose that $K_j > K_i$. For any indices $1 \le j_1, \cdots, j_\ell \le m$ we shall define the set $\{j_1, \cdots, j_\ell\}$ of δ-pseudo-orbits $\omega = (x_0, \cdots, x_n)$ with $\delta \le \delta(\theta_0)$ saying that $\omega \in \{j_1, \cdots, j_\ell\}$ if and only if for all $i = 1, \cdots, n$,

$$x_i \in \Big(\bigcup_{r=1}^{\ell} U^G_{\delta(\theta_0)}(K_{j_r})\Big) \cup \Big(\bar{G} \setminus \bigcup_{q=1}^{m} U^G_{\delta(\theta_0)}(K_q)\Big)$$

and there exist indices

$$k_0(\omega) = 0 \le i_1(\omega) \le k_1(\omega) \le \cdots \le i_\ell(\omega) \le k_\ell(\omega) \le i_{\ell+1}(\omega) = n$$

such that for $q = 1, \cdots, \ell$,

$$i_q(\omega) = \inf\{r \ge k_{q-1}(\omega) : x_r \in U^G_{\delta(\theta_0)}(K_{j_q})\},$$

$$k_q(\omega) = \inf\{r > i_q(\omega) : x_r \notin U^G_{\theta_0}(K_{j_q})\},$$

and if $k_\ell(\omega)$ is not defined by the last relation, i.e., if $x_r \in U^G_{\theta_0}(K_{j_\ell})$ for all $r \ge i_\ell(\omega)$ then we put $k_\ell(\omega) = n$. From Lemma 2.1 it follows that if $\{j_1, \cdots, j_\ell\}$ is not empty then $K_{j_\ell} > K_{j_{\ell-1}} > \cdots > K_{j_1}$ and all these compacts are different. Furthermore, by Lemma 2.2 if $\delta \le \frac{1}{2}\delta_0 C_2^{-(T_0+1)}$ then $i_{q+1}(\omega) - k_q(x) \le T_0$.

To prove Corollary 2.1 it suffices to consider the case when for any $\delta > 0$ there exists a δ-pseudo-orbit $\omega^{(\delta)} = (x_0^{(\delta)}, \cdots, x_{n(\delta)}^{(\delta)})$ such that $x_0^{(\delta)} \in K_i$ and $x_{n(\delta)}^{(\delta)} \in K_j$. Taking into account the above arguments it is easy to see that one can choose a sequence $\delta_r \longrightarrow 0$ as $r \longrightarrow \infty$ and indices $j_1 = i, j_2, \cdots, j_\ell = j$ such that for r big enough $\omega^{(\delta_r)} \in \{j_1, \cdots, j_\ell\}$ and there exist limits

$$z_q = \lim_{r \to \infty} x^{(\delta_r)}_{k_q(\omega^{(\delta_r)})} \quad \text{for all } q = 1, \cdots, \ell-1.$$

Since the limit set of the dynamical system F^t in each $U^G_{\theta_0}(K_{j_q})$ must be contained in K_{j_q}, then either dist $(F^{-t}z_q, K_{j_q}) \longrightarrow 0$ as

$t \longrightarrow \infty$ or there exists a positive $t_q < \infty$ such that $F^{-t_q} z_q = z_{q-1}$. Similarly, either dist $(F^t z_q, K_{j_{q+1}}) \longrightarrow 0$ as $t \longrightarrow \infty$ or one can find a positive $\tilde{t}_q < \infty$ such that $F^{\tilde{t}_q} z_q = z_{q+1}$. Since $x_0^{(\delta_r)} \in K_i$ and $x_{n(\delta_r)}^{(\delta_r)} \in K_j$ for the whole sequence $\delta_r \longrightarrow 0$ then it follows that dist $(F^{-t} z_1, K_{j_1})$ + dist $(F^t z_{\ell-1}, K_{j_\ell}) \longrightarrow 0$ as $t \longrightarrow \infty$. Now put $y_1 = z_1$, $K_{r_1} = K_{j_1}$, and then define successively $y_{k+1} = z_q$, $K_{r_{k+1}} = K_{j_q}$ provided dist $(F^t y_k, K_{j_q}) \longrightarrow 0$ as $t \longrightarrow \infty$. It is easy to see that the points $\{y_k\}$ and the compacts $\{K_{r_k}\}$ satisfy (2.21). □

Next, we come back to the proof of Theorem 2.1. Put $\delta_2 = \frac{1}{2}\delta_o C_2^{-(T_o+1)}$. It follows from Lemmas 2.1 and 2.2 that any δ_2-pseudo-orbit $\omega = (x_0, \cdots, x_n)$ belongs to a set $\{j_1, \cdots, j_n\}$ with some $\ell \le m$ and the corresponding indices $k_{q-1}(\omega) \le i_q(\omega) \le k_q(\omega) \le n$, $q = 1, \cdots, \ell$ satisfying $\sum_{1 \le q \le \ell} (k_q(\omega) - i_q(\omega)) \ge n - T_o(m+1)$. Thus for any $x \in G$ and an integer $n > 0$ we can write

$$I_1^\epsilon(\delta_2, n, x) \overset{\text{def}}{=} P_x^\epsilon\{\text{dist}(F^1 X_i^\epsilon, X_{i+1}^\epsilon) < \delta_2 \text{ for all } i = 0, \cdots, n-1 \text{ and } X_t^\epsilon \in G \text{ for all } t \in [0,n]\} \tag{2.22}$$

$$\le \sum_{1 \le \ell \le m} \sum_{j_1, \cdots, j_\ell} \sum_{i_1 \le k_1 \le \cdots \le i_\ell \le k_\ell} \tilde{I}_x^\epsilon(j_1, \cdots, j_\ell; i_1, \cdots, i_\ell; k_1, \cdots, k_\ell)$$

where

$$\tilde{I}^{\epsilon}_x(j_1,\cdots,j_\ell;i_1,\cdots,i_\ell;k_1,\cdots k_\ell)$$

$$= P^{\epsilon}_x\{X^{\epsilon}_r \in U_{\theta_o}(K_{j_q}) \cap G \text{ for all } r=i_q,\cdots,k_q-1$$
and all $q=1,\cdots,\ell\}$,
the second sum in (2.22) is taken over $j_1,\cdots,j_\ell$ such that $\{j_1,\cdots,j_\ell\} \neq \Phi$, and the third sum is taken over $i_1 \le k_1 \le \cdots \le i_\ell \le k_\ell$ satisfying

$$\sum_{1\le q\le \ell} (k_q-i_q) \ge n_o - T_o(m+1). \tag{2.23}$$

It is clear that the total number of elements in the sum in the right hand side of (2.22) does not exceed $m^m n^{2m}$. Hence this sum can be estimated by $m^m n^{2m}$ - times the maximal element in the sum, i.e.,

$$I^{\epsilon}_1(\delta_2,n,x) \tag{2.24}$$

$$\le m^m n^{2m} \max_{\ell\le\nu;\, j_1,\cdots,j_\ell;\, i_1\le k_1\le\cdots\le i_\ell\le k_\ell} \tilde{I}^{\epsilon}_x(j_1,\cdots,j_\ell;i_1,\cdots i_\ell;k_1,\cdots,k_\ell)$$

where the maximum is taken over the same set of indices as in the sum (2.22). By the strong Markov property of the process X^{ϵ}_t if follows that

$$\tilde{I}^{\epsilon}_x(j_1,\cdots j_\ell;i_1,\cdots,i_\ell;k_1,\cdots,k_\ell) \tag{2.25}$$
$$= E^{\epsilon}_x E^{\epsilon}_{X^{\epsilon}_{i_1}} \chi_{X^{\epsilon}_r \in U_{\theta_o}(K_{j_1}) \cap G \text{ for } r=0,\cdots,k_1-i_1-1}$$
$$\times\cdots\times E^{\epsilon}_{X^{\epsilon}_{i_\ell}} \chi_{X^{\epsilon}_r \in U_{\theta_o}(K_{j_1}) \cap G \text{ for } r=0,\cdots,k_\ell-i_\ell-1} \le$$
$$\le \prod_{1\le q\le \ell} \sup_{y\in U_{\theta_o}(K_{j_q})} I^{\epsilon}_3(j_q,k_q-i_q y)$$

where

$$I_3^\epsilon(j,r,y)=P_y^\epsilon\{X_s^\epsilon \in U_{\theta_o}(K_j)\cap G \text{ for all } s=1,\cdots,r-1\} \tag{2.26}$$

$$= \int_{U_{\theta_o}(K_j)} \cdots \int_{U_{\theta_o}(K_j)} p_G^\epsilon(1,y,z_1)p_G^\epsilon(1,z_1,z_2)\cdots p_G^\epsilon(1,z_{r-1},z_r)dz_1\cdots dz_r$$

and Π denotes the product.

Using again the strong Markov property of the process X_t^ϵ we derive that for any $v,w \in V_j\cap G$,

$$p_G^\epsilon(1,v,w) = p_{V_j\cap G}^\epsilon(1,v,w) \tag{2.27}$$

$$+ E_v^\epsilon \chi_{\tau(V_j)<1}\, p_G^\epsilon(1-\tau(V_j),X_{\tau(V_j)}^\epsilon,w) \quad .$$

Notice, that if $v \in U_{\theta_o}(K_j)$ then in view of (2.16) and (2.17),

$$\inf_{0\leq t\leq 1} \text{dist } (F^t v,\partial V_j) \geq C_2\theta_o. \tag{2.28}$$

This together with the large deviations estimates from §1 and §2 of Wentzell and Freidlin [WF] yield

$$E_v^\epsilon\, \chi_{\tau(V_j)<1} = P_v^\epsilon\{\tau(V_j) < 1\} \tag{2.29}$$

$$\leq P_v^\epsilon\{\sup_{0\leq t\leq 1} \text{dist } (X_t^\epsilon,F^t v) \geq C_2\theta_o\}$$

$$\leq C_3 \exp(-\delta_3\, \epsilon^{-2})$$

provided $v \in U_{\theta_o}(K_j)$ where $C_3,\ \delta_3 > 0$ are independent of $v,\epsilon>0$, and $j = 1,\cdots,m$.

By (II.7.25), (2.28), and (2.29) we derive from (2.27) that

$$p_G^\epsilon(1,v,w) \leq p_{V_j\cap G}^\epsilon(1,v,w) + C_4\epsilon^{-\upsilon}\exp(-\delta_3\epsilon^{-2}) \tag{2.30}$$

provided $v,w \in U_{\theta_o}(K_j)$ where $C_4 > 0$ is the product of constants from the right hand sides of (II.7.25) and (2.29). Substituting (2.30) into (2.26) we obtain

$$I_3^{\epsilon}(j,r,y) \le C_4 r \epsilon^{-\upsilon} \exp(-\delta_3 \epsilon^{-2}) \tag{2.31}$$
$$+\int_{U_{\theta_o}(K_j)} \cdots \int_{U_{\theta_o}(K_j)} p^{\epsilon}_{V_j \cap G}(1,y,z_1) \cdots p^{\epsilon}_{V_j \cap G}(1,z_{r-1},z_r) dz_1 \cdots dz_r$$
$$\le C_4 r\, \epsilon^{-\upsilon} \exp(-\delta_3 \epsilon^{-2}) + \Phi^{\epsilon}(r, V_j \cap G)$$

whereas $y \in U_{\theta_o}(K_j)$. By (2.10) we have also for any $t > 0$,

$$\Phi^{\epsilon}(r,V_j) \le (\Phi^{\epsilon}(t, V_j \cap G))^{[\frac{r}{t}]} \tag{2.32}$$

where, again, $[\cdot]$ denotes the integral part.

Collecting (2.13), (2.24), (2.25), (2.31), (2.32), and taking into account (2.23) we derive

$$\Phi^{\epsilon}(n,G) \le m^m n^{2m} (\max_{1 \le j \le m} \Phi^{\epsilon}(t,V_j))^{(n-T_o(m+1)-tm)t^{-1}} \tag{2.33}$$
$$+ m^m n^{2m}((C_4 n \epsilon^{-\upsilon} \exp(-\delta_3 \epsilon^{-2}) + 1)^m - 1)$$
$$+ C_1 \epsilon^{-\upsilon} n \exp(-\beta_1 \delta_2^2 \epsilon^{-2}).$$

Now if (2.4) is true for some $t = t(\epsilon,\delta) \le e^{-2(1-\beta_o)}$ then taking $n = n(\epsilon) = [\epsilon^{-2+\beta_o}]$ we obtain for any ϵ small enough that

$$\Phi^{\epsilon}(n(\epsilon),G) \le 2m^m (n(\epsilon))^{2m} \tag{2.34}$$
$$\times \exp((\delta + \max_{1 \le j \le m} \lambda(K_j))(n(\epsilon) - T_o(m+1) - t(\epsilon,\delta))).$$

By (2.11) and the choice of $n(\epsilon)$ and $t(\epsilon,\delta)$ this implies

$$\lambda^{\epsilon}(G) \leq \max_{1\leq j\leq m} \lambda(K_j) + \delta + C_5(\epsilon^{2-\beta_0} + \epsilon^{\beta_0})\log(\tfrac{1}{\epsilon}) \qquad (2.35)$$

for some $C_5 > 0$ independent of ϵ.
Letting in (2.35) first $\epsilon \longrightarrow 0$ and then $\delta \longrightarrow 0$ we obtain

$$\limsup_{\epsilon\to 0} \lambda^{\epsilon}(G) \leq \max_{1\leq j\leq m} \lambda(K_j)$$

which together with (2.9) proves Theorem 2.1. □

Remark 2.5. The upper bound (2.4) with the restriction (2.3) is necessary to pass from (2.33) to (2.34) in order to avoid the possibility when $\Phi^{\epsilon}(t,V_j)$ is of the same order as correction terms which we want to eliminate.

In Remark 2.2 we discussed how to find the numbers $\lambda(K_i)$ for certain types of compacts K_i lying strictly inside of G. Next, we are going to consider the situation when K_i is an F^t-invariant connected component Γ of the boundary ∂G of G. Thus Γ is a closed smooth surface of the codimension one. It is easy to see that one can choose an open neighborhood V of Γ in M such that any point $x \in V$ has a unique representation in local coordinates $x = \gamma(x) + \rho(x)n(x)$ where $\gamma(x) \in \Gamma$, $|\rho(x)| = \mathrm{dist}(x,\Gamma)$, and $n(x) = n(\gamma(x))$ is the interior unit normal to Γ in the sense that it points out into the interior of G, and so $\rho(x) > 0$ if $x \in V \cap G$. We can characterize any point $x \in V$ by the pair $(\gamma(x),\rho(x))$. Then the normal to Γ component $q(x)$ of the vector field B(x) satisfies

$$\frac{dq(F^t x)}{dt} = q(F^t x) = q(\gamma(F^t x),\rho(F^t x)). \qquad (2.36)$$

For each $\gamma \in \Gamma$ define

$$\alpha(\gamma) = \left.\frac{\partial q(\gamma,\rho)}{\partial \rho}\right|_{\rho=0} \tag{2.37}$$

Then for each $x \in V$ one can write

$$q(x) = \alpha(\gamma(x))\rho(x) + \Psi(x)\rho^2(x) \tag{2.38}$$

where Ψ is a bounded function in V.

We shall need

Assumption 2.2. Uniformly in $\gamma \in \Gamma$ the limit

$$\lim_{t\to\infty} \frac{1}{t}\int_0^t \alpha(F^u\gamma)du = \alpha_o \tag{2.39}$$

exists and it is independent of γ.

Remark 2.6. The above assumption is satisfied if the dynamical system F^t restricted to the F^t-invariant surface Γ is uniquely ergodic, i.e., it has a unique invariant measure on Γ. This follows from the continuous time version of Theorem 6.19 in Walters [Wa].

The following result was established by the author together with A. Eizenberg in [EK].

Theorem 2.2. *Let Γ be an F^t-invariant connected component of the boundary ∂G and let V be an open neighborhood of Γ with a smooth boundary ∂V such that $V \cup \partial V$ contains no closed F^t-invariant set except for Γ. Suppose that Assumption 2.2 holds true and put $V^G = V \cap G$.*
(i) If $\alpha_o < 0$ (the case of an attracting boundary) then the limit

$$\lambda(\Gamma) = \lim_{\varepsilon\to 0}\lim_{t\to\infty} \frac{1}{t}\log \Phi^{\varepsilon}(t,V^G) \tag{2.40}$$

exists and $\lambda(\Gamma) = \alpha_o$;
(ii) If $\alpha(\gamma) \equiv 0$ on Γ (the case of a neutral boundary) then the limit (2.40) exists and $\lambda(\Gamma) = 0$;

(iii) *If* $\alpha_0 > 0$ *(the case of a repulsing boundary) and, in addition, the dinamical system* F^t *restricted to* Γ *has an invariant measure on* Γ *possessing a smooth positive density with respect to the induced Riemannian volume on* Γ*, then* $\lambda(\Gamma) = -2\alpha_0$.

Furthermore, in the above cases, for any $\delta, \beta > 0$,

$$\Phi^{\epsilon}(t, V^G) \leq \exp((\lambda(\Gamma) + \delta)t) \tag{2.41}$$

provided $\epsilon \leq \epsilon(\delta)$ *and* $t \geq (\log \frac{1}{\epsilon})^{1+\beta}$*, and so* Γ *and* V^G *can play the role of a pair* K_i *and* V_i *in Assumption 2.1.*

The proof of this theorem is long, technical, and it involves extensive estimates of solutions of corresponding stochastic differential equations. By this reason we do not exhibit it here and refer the reader to [EK].

Remark 2.7. One of examples we have in mind which satisfies the assumptions of Theorem 2.2 is the case when the flow F^t on Γ is diffeomorphically conjugate to an irrational rotation on an $(\upsilon-1)$-dimensional torus (see Example I.1.2). According to Lind [Li] this will be the case if F^t restricted to Γ is a dynamical system with a discrete (pure-point) spectrum and smooth eigenfunctions.

Remark 2.8. It would be interesting to prove (ii) under the weaker condition $\alpha_0=0$. In (iii) we need a smooth invariant measure since we derive (iii) from (i) passing to an adjoint to L^{ϵ} operator. The use of direct probabilistic estimates may eliminate this condition. The adjoint operator method works in some other cases, as well. Suppose that Γ is an F^t-invariant $(\upsilon-1)$-dimensional repulsing surface in the sense of the assertion (iii) of Theorem 2.2 lying in G. Assume that there exists an open set $V \subset G$ with a smooth boundary which does not contains any other F^t-invariant set except for Γ. The adjoint to L^{ϵ} operator having the same principal eigenvalue has one part which corresponds to the attracting surface case, which according to Remark 2.3 gives zero principal eigenvalue, and it has another part which yields $-\alpha_o$. Thus in this case $\lim_{\epsilon\to 0} \lambda^{\epsilon}(V) = -\alpha_o$. A similar result holds true if Γ is not necessarily of the co-dimension one, in particular, this method works for certain types of normally hyperbolic manifolds defined by Hirsh, Pugh, and Shub [HPS].

If a connected component Γ of the boundary ∂G of G is not F^t-invariant, but Γ contains F^t-invariant subsets then, in general, the situation becomes more complicated. Still, combining the results described in Remark 2.2 with Theorem 2.2 one can treat some of these cases.

Assume, for example, that $\mathcal{O} \in \Gamma$ is a fixed point of the dynamical system F^t isolated from the rest of the limit set. Then we can write locally

$$B(x) = \Pi\,(x-\mathcal{O}) + O((\mathrm{dist}(x,\mathcal{O}))^2) \tag{2.42}$$

where Π is a matrix. Suppose that Π has an eigendirection ξ which is transversal to Γ at $\mathcal{O}$. Then any vector x can be uniquely represented using local coordinates as $x = x_{\xi} + x_{\Gamma}$ where $x_{\xi} \in \xi$ and x_{Γ} belongs to the tangent hyperplane $T_{\mathcal{O}}\Gamma$ to Γ at $\mathcal{O}$. Let V be a small neighborhood of $\mathcal{O}$ with a smooth boundary and $V^G = V \cap G$. It is not difficult to understand that the study of the exit time from V^G for the process X_t^{ϵ} needed to determine $\lambda(\mathcal{O})$ by (2.2), can be carried out, actually, independently for projections of X_t^{ϵ} into ξ and $T_{\mathcal{O}}\Gamma$. The projection into ξ can be treated by means of the one-dimensional version of Theorem 2.2 and the projection into $T_{\mathcal{O}}\Gamma$ is being studied according to Remark 2.2. Thus, if γ_1 is an eigenvalue corresponding to ξ and $\gamma_2, \cdots, \gamma_{\upsilon}$ are other eigenvalues of Π then

$$\lambda(\mathcal{O}) = -(|\mathrm{Re}\gamma_1| + \sum_{i=1}^{\nu} \max(0, \mathrm{Re}\gamma_i)). \qquad (2.43)$$

Combining Theorem II.7.1 with the one-dimensional version of Theorem 2.2 one can obtain corresponding results also for other types of F^t-invariant subsets of Γ. If the matrix Π does not have an eigendirection transversal to Γ at $\mathcal{O}$ then the corresponding asymptotics can not be derived readily in the way described above. Still, one can solve the problem using the Gaussian approximation $e^{t\Pi}(x+\epsilon\int_0^t e^{-u\Pi}\sigma(\mathcal{O})dw_t)$ of the process X_t^ϵ near $\mathcal{O}$.

Next, we shall exhibit an example with a quite different asymptotical behavior of the principal eigenvalue $\lambda^\epsilon(G)$. The orbits of the dynamical system F^t in our example are indicated on the following picture by thin lines and G is a ring between two connected components Γ_1 and Γ_2 of the boundary ∂G which are drawn as boldface circles.

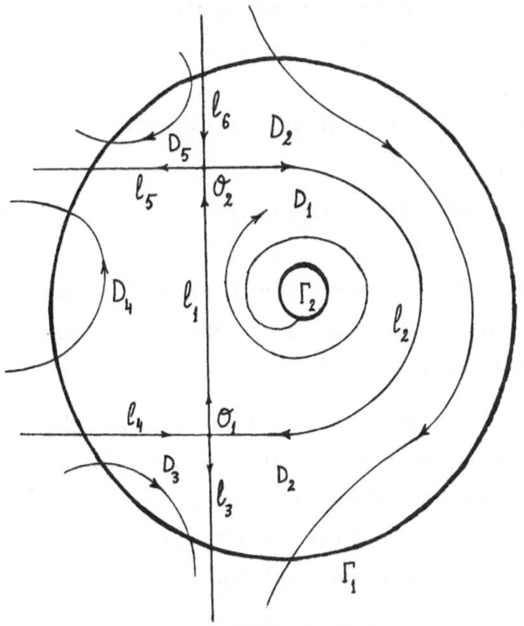

Figure 2.1

The limit set of the dynamical system F^t in $\bar{G}$ consists of the loop $\mathcal{L}$ which contain two fixed points O_1 and O_2 of the saddle type, and the orbits

$\ell_1 = \{z: \ F^t z \longrightarrow O_2$ and $F^{-t} z \longrightarrow O_1$ and $t \longrightarrow \infty\}$,

$\ell_2 = \{z: \ F^t z \longrightarrow O_1$ and $F^{-t} z \longrightarrow O_2$ and $t \longrightarrow \infty\}$.

All orbits of F^t entering G through Γ_2 approach eventually the loop $\mathcal{L}$. On the other hand, all orbits in the exterior of the loop $\mathcal{L}$ except for the stable curve of O_1 leave G through Γ_1. Near the fixed points O_1 and O_2 we have the representation (2.42) with some matrices Π_1 and Π_2, respectively. Let $\gamma_1^{(1)}$, $\gamma_2^{(1)}$ and $\gamma_1^{(2)}$, $\gamma_2^{(2)}$ be the eigenvalues of Π_1 and Π_2, correspondingly, such that $Re\gamma_1^{(1)} < 0 < Re\gamma_2^{(1)}$ and $Re\gamma_1^{(2)} < 0 < Re\gamma_2^{(2)}$. Our picture corresponds to an attracting loop $\mathcal{L}$ which will be satisfied if $|Re\gamma_1^{(1)}| > Re\gamma_2^{(1)}$ and $|Re\gamma_1^{(2)}| > Re\gamma_2^{(2)}$.

The whole loop $\mathcal{L}$ is the single equivalence class, and so we cannot employ Theorem 2.1 here, i.e., the study of exit times from small neighborhoods of the fixed points O_1 and O_2 (which can be done according to Remark 2.2) does not help here. This can be seen also from the following result.

Proposition 2.3. *In the above example,*

$$C^{-1}|\log\varepsilon|^{-1} \le |\lambda^{\varepsilon}(G)| \le C|\log\varepsilon|^{-1} \tag{2.44}$$

for some constant $C > 0$ *independent of* $\varepsilon > 0$ *small enough.*

Proof. We shall outline only the main idea of the proof which can be found in Eizenberg and Kifer [EK]. Each time when the process X_t^{ε} arrives to the ε-neighborhood of one of the points O_1 and O_2 it proceeds to O_2 and O_1, respectively, rather then exits from G with probability sandwiched between two positive constants $\rho_1 > \rho_2$ independent of ε. It can be explained roughly in the following way. Introduce the subdomains $D_i \subset G$, $i=1,\cdots,5$ bounded by stable and unstable curves ℓ_i, $i=1,\cdots,6$ and by the boundaries Γ_1, Γ_2 as it is pointed out on the above picture. Then X_t^ϵ proceeds from $\mathcal{O}_1$ to $\mathcal{O}_2$ and from $\mathcal{O}_2$ to $\mathcal{O}_1$ if it stays in D_1. By the symmetry the probability for X_t^ϵ to stay in D_1 if X_t^ϵ starts extremely close to $\mathcal{O}_1$ or $\mathcal{O}_2$ is about $\frac{1}{4}$.

On the other hand, according to Remark 2.4 the process X_t^ϵ exits from ϵ-neighborhoods of the points $\mathcal{O}_1$ and $\mathcal{O}_2$ for the time of order $|\log\epsilon|$. This together with the above discussion lead to the conclusion that

$$\tilde{C}^{-1} \rho_2^{t/|\log\epsilon|} \leq P_x^\epsilon\{\tau(G)>t\} \leq \tilde{C} \rho_1^{t/|\log\epsilon|} \qquad (2.45)$$

which yields (2.44). □

Remark that the speed of convergence of $\lambda^\epsilon(G)$ to zero represented by (2.44) is new since in all previously known cases this speed of convergence was either polinomial in ϵ or exponentially fast in $(-\epsilon^{-1})$ as in §7 of Chapter 6 from Freidlin and Wentzell [FW].

3.3 Random pertubations and the spectrum

In this section following [Ki4] we shall study the asymptotical behavior as $\epsilon \longrightarrow 0$ of the whole spectrum of the operator L^ϵ.

A dynamical system F^t on a space **M** is said to have a pure-point (discrete) spectrum if there exists a set of real real numbers $\{\gamma_k\}$ and an orthonormal basis $\{g_k\}$ in L^2-space with respect to F^t-invariant measure such that

$$g_k(F^t x) = e^{i\gamma_k t} g_k(x), \quad i = \sqrt{-1}. \qquad (3.1)$$

If M is a smooth manifold and $\{g_k\}$ are smooth functions then according to Lind [Li] there exists a diffeomorphism S which maps the manifold M onto the υ-dimensional torus $\mathcal{T}^{\upsilon}$ and

$$SF^t = R^tS \tag{3.2}$$

where R^t is the one-parameter group of rotations of $\mathcal{T}^{\upsilon}$ (the quasiperiodic motion on $\mathcal{T}^{\upsilon}$), i.e.,

$$\frac{dR^tx}{dt} = B = (\omega_1 \frac{\partial}{\partial\varphi_1}, \cdots, \omega_{\upsilon} \frac{\partial}{\partial\varphi_{\upsilon}}) \tag{3.3}$$

where the right hand side of (3.3) is the constant vector field on $\mathcal{T}^{\upsilon}$, $\omega = (\omega_1, \cdots, \omega_{\upsilon})$ are called frequencies, and $\varphi = (\varphi_1, \cdots, \varphi_{\upsilon})$ are cyclic coordinates. This justifies partially that we shall restrict ourselves to random perturbations of the quasiperiodic motion only.

Consider an elliptic operator

$$L = \frac{1}{2} \sum_{1 \leq k,\ell \leq \upsilon} a^{k\ell}(\varphi)\frac{\partial^2}{\partial\varphi_k\partial\varphi_\ell} + \sum_{1 \leq k \leq \upsilon} b^k(\varphi)\frac{\partial}{\partial\varphi_k},$$

$$\varphi = (\varphi_1, \cdots, \varphi_{\upsilon}) \tag{3.4}$$

on $\mathcal{T}^{\upsilon}$ with 2π-periodic coefficients where the matrix $A(\varphi) = (a^{k\ell}(\varphi))$ is supposed to be uniformly positive definite. The random perturbations X_t^{ϵ} are generated by the operators $L^{\epsilon} = \epsilon^2L+B$ in the sense explained in Section 3.1. The corresponding transition operator P_t^{ϵ} is completely continuous, and so its spectrum consists of at most countable set of numbers which may accumulate only to zero. There is one-to-one correspondence between the spectrum of

L^ϵ consisting of eigenvalues $\{\lambda_m^\epsilon\}$ and the spectrum of P_t^ϵ consisting of eigenvalues $\{e^{\lambda_m^\epsilon t}\}$. Our approach will be purely analytic, and so we shall deal only with the operator L^ϵ rather than refering to probabilistic properties of the process X_t^ϵ or its transition operator P_t^ϵ. The coefficients of L may be even complex.

Let a function $v(\varphi)$ on $\mathcal{T}^\nu$ has the Fourier expansion

$$v(\varphi) = \sum_q v_q e^{i(q,\varphi)}$$

where $q = (q_1,\cdots,q_\nu)$ is a set of integers and $(q,\varphi) = q_1\varphi_1 + \cdots + q_\nu\varphi_\nu$. The family of norms $\|\cdot\|_\rho$, $\rho \geq 0$ is defined by the formula

$$\|v\|_\rho^2 = |v_0|^2 + \sum_q |v_q|^2 |q|^{2\rho} \tag{3.5}$$

where $|q|^{2\rho} = (q_1^2 + \cdots + q_\nu^2)^\rho$, $v_0 = v_{(0,\cdots,0)}$, and $0^0 = 1$. The closure of the set of all trigonometrical polynomials with respect to the norm $\|\cdot\|_\rho$ is called the Sobolev space H^ρ. The inner product in this space is given by the formula

$$(v,w)_\rho = v_0\bar{w}_0 + \sum_q v_q\bar{w}_q |q|^{2\rho} \tag{3.6}$$

where $\bar{w}$ is the complex conjugate of w (sometimes $(v,w)_0$ will be denoted simply by (v,w)).

The operator B has eigenvalues of the form $\lambda_m = i(m,\omega)$, where $m = (m_1,\cdots,m_\nu)$ and $\omega = (\omega_1,\cdots,\omega_\nu)$, and corresponding eigenfunctions $r_m(\varphi) = \exp\{i(m,\varphi)\}$, $\varphi = (\varphi_1,\cdots,\varphi_\nu)$. The frequencies $\omega = (\omega_1,\cdots\omega_\nu)$ are called rationally independent, if the equality $(m,\omega) = k$ for an integer k and an integer vector $m = (m_1,\cdots,m_\nu)$ can be fulfilled only if $k = m_1 = \cdots = m_\nu = 0$.

Let the operator L defined by (3.4) has complex periodic coefficients $a^{k\ell}(\varphi)$ and $b^k(\varphi)$, $k,\ell = 1,\cdots,n$ belonging to the space H^α, $\alpha > 0$, and the frequencies $(\omega_1,\cdots,\omega_\nu)$ be rationally independent. We assume that the operator L is uniformly strongly elliptic, i.e., the matrix $A(\varphi) = (a^{k\ell}(\varphi))$ satisfies

$$\gamma^{-1}(x,x) \geq \mathrm{Re}(A(\varphi)x,x) \geq \gamma(x,x) \qquad (3.7)$$

for some constant $\gamma > 0$, $\gamma \leq 1$ independent of $x = (x_1,\cdots x_\nu)$ and $\varphi = (\varphi_1,\cdots\varphi_\nu)$, where Re denotes the real part and $(x,y) = \sum_\ell x_\ell \overline{y}_\ell$.

The main result of this section is the following

Theorem 3.1. *(i) For any set of integers* $m = (m_1,\cdots,m_n)$ *and* $\epsilon > 0$ *small enough there exist an eigenfunction* $r_m^\epsilon \in H^{2+\alpha}$ *and an eigenvalue* λ_m^ϵ *of the operator* L^ϵ *such that for* $0 < \beta < \alpha$ *one has*

$$(r_m^\epsilon, r_m) = 1,\ \|r_m^\epsilon\|_{2+\alpha} \leq C_m,\ \|r_m^\epsilon - r_m\|_{2+\alpha-\beta} \leq C_{m,\beta}(\epsilon), \qquad (3.8)$$

$$|\lambda_m^\epsilon - \lambda_m| \leq C_m\epsilon^2, \qquad (3.9)$$

where C_m *depends on* $|m|$ *and the constant of ellipticity* γ *of the operator* L, $C_{m,\beta}(\epsilon) \longrightarrow 0$ *as* $\epsilon \longrightarrow 0$ *and the speed of this convergence depends on approximative properties of the frequencies* $(\omega_1,\cdots,\omega_\nu)$. *Provided* $\epsilon > 0$ *is small enough, the function* r_m^ϵ *and the number* λ_m^ϵ *are defined uniquely by these conditions.*

(ii) There exist constants $C^{(1)}$, $C^{(2)} > 0$ *and a positive integer* Q, *such that for an arbitrary eigenvalue* λ^ϵ *of* L^ϵ

satisfying the condition $\mathrm{Re}\lambda^{\epsilon} \geq -C^{(1)}\epsilon^{2}$, *there exists an integer vector* $q = (q_1, \cdots, q_{\nu})$ *such that* $|q| = (q_1^2 + \cdots, q_{\nu}^2)^{1/2} \leq Q$ and

$$|\mathrm{Im}\lambda^{\epsilon} - (q, \omega)| \leq C^{(2)}\epsilon^{2}, \tag{3.10}$$

where Im *denotes the imaginary part.* Among such *eigenvalues there exist* μ *numbers* $\lambda^{\epsilon}_{m_1}, \cdots, \lambda^{\epsilon}_{m_{\nu}}$ *satisfying the inequalities*

$$|\lambda^{\epsilon}_{m_k} - \omega_k| \leq C^{(2)}\epsilon^{2}, \quad k = 1, \cdots, \nu. \tag{3.11}$$

(*This gives an approximation method for the determination of generators of the spectrum of* B *by the spectrum of* L^{ϵ}).

Proof. If an eigenfunction r^{ϵ}_m and an eigenvalue λ^{ϵ}_m of L^{ϵ} satisfy the conditions

$$L^{\epsilon}r^{\epsilon}_m = \lambda^{\epsilon}_m r^{\epsilon}_m \text{ and } (r^{\epsilon}_m, r_m) = 1 \tag{3.12}$$

then

$$\lambda^{\epsilon}_m = i(m, \omega) + \epsilon^{2}(Lr^{\epsilon}_m, r_m) \tag{3.13}$$

and

$$\epsilon^{2}Lr^{\epsilon}_m + Br^{\epsilon}_m - i(m,\omega)r^{\epsilon}_m = \epsilon^{2}(Lr^{\epsilon}_m, r_m)r^{\epsilon}_m. \tag{3.14}$$

Introduce the operator Φ^{ϵ}_m: $H^{2+\alpha} \longrightarrow H^{2+\alpha}$ acting by the formula $\Phi^{\epsilon}_m f = q$, where

$$\epsilon^{2}C\Delta g + Bg - i(m,\omega)\, g = \epsilon^{2}(C\Delta - L)\, f + \epsilon^{2}(Lf, r_m)f, \tag{3.15}$$

$\Delta = \sum_{k \leq \nu} \frac{\partial^2}{\partial \varphi_k^2}$ is the Laplacian and $C > 0$ is a constant which will be chosen later.

It is clear that r_m^ϵ is a fixed point of the operator Φ_m^ϵ and if $|m| \neq 0$ then the following implication holds true

$$\text{if } (f, r_m) = 1 \text{ then } (\Phi_m^\epsilon f, r_m) = 1. \tag{3.16}$$

Let $G_m^\epsilon\colon\ H^\alpha \longrightarrow H^{2+\alpha}$ be the linear operator defined by

$$G_m^\epsilon\, e^{i(q,\varphi)} = \begin{cases} (-\epsilon^2 C|q|^2 + i(q-m,\omega))^{-1}\ e^{i(q,\varphi)} \text{if } q \neq m \\ 0 \text{ if } q = m, \end{cases} \tag{3.17}$$

i.e., G_m^ϵ equals $(\epsilon^2 C\Delta + B - i(m,\omega))^{-1}$ on a subspace orthogonal to r_m and $G_m r_m = 0$. If $(f, r_m) = 1$ then in view of (3.15) - (3.17),

$$\Phi_m^\epsilon f = r_m + \epsilon^2 G_m^\epsilon (C\Delta - L)\, f + \epsilon^2 (Lf, r_m) G_m^\epsilon f \tag{3.18}$$

Our goal will be to show that the operator Φ_m^ϵ turns out to be a contraction in some ball in $H^{2+\alpha}$. Thus Φ_m^ϵ has a unique fixed point r_m^ϵ there which satisfies (3.14), and so (3.12) - (3.13) hold true, as well.

Let

$$A(\varphi) = \sum_q A_q e^{i(q,\varphi)} \quad \text{and} \quad b(\varphi) = \sum_q b_q\ e^{i(q,\varphi)}$$

be the Fourier expansions for the coefficients $A(\varphi) = (a^{k\ell}(\varphi))$ and $b(\varphi) = (b^k(\varphi))$ of the operator L defined by (3.4), where $\{A_q\}$ are matrices and $\{b_q\}$ are vectors. Since these coefficients belong to the space H^α then we can choose $N > 2$ such that

$$\|Lf - L^{(N)}f\|_\alpha \leq \frac{\gamma}{32} \|f\|_{2+\alpha} \qquad (3.19)$$

where γ is the same as in (3.7), and the operator $L^{(N)}$ has the coefficients $A^{(N)}(\varphi) = (a_N^{k\ell}(\varphi))$ and $b^{(N)}(\varphi) = (b_N^k(\varphi))$ written in place of $A(\varphi)$ and $b(\varphi)$ in (3.4) and having the form

$$A^{(N)}(\varphi) = \sum_{q:|q|\leq N} A_q e^{i(q,\varphi)} \text{ and}$$

$$b^{(N)}(\varphi) = \sum_{q:|q|\leq N} b_q e^{i(q,\varphi)}. \qquad (3.20)$$

If N is large enough then in view of (3.7) we have

$$2\gamma^{-1}(x,x) \geq \mathrm{Re}(A^{(N)}(\varphi)x,x) \geq \frac{\gamma}{2}(x,x). \qquad (3.21)$$

Now we shall fix some N satisfying (3.19) and (3.21). Put
$z = 1 + \nu(\max_{k,\ell\leq\nu} \|a^{k\ell}(\varphi)\|_\alpha + \max_{k\leq\nu} \|b^k(\varphi)\|_\alpha)$.
From the definition (3.5) of Sobolev's norms it follows that

$$\|Lf\|_\alpha \leq Z\|f\|_{2+\alpha}, \quad \|L^{(N)}f\|_\alpha \leq Z\|f\|_{2+\alpha} \text{ and}$$

$$|(Lf,r_m)| \leq Z\|f\|_2. \qquad (3.22)$$

Next, we shall rewrite (3.18) in the form

$$\Phi_m^\epsilon f = r_m + \epsilon^2 G_m^\epsilon (C\Delta - L^{(N)})f + \epsilon^2 G_m^\epsilon (L^{(N)} - L) f \qquad (3.23)$$

$$+ \epsilon^2 (Lf,r_m) G_m^\epsilon f.$$

From (3.17) and (3.19) we derive that

$$\epsilon^2 \| G_m^\epsilon (L-L^{(N)}) f \|_{2+\alpha} \leq \frac{\gamma}{32C} \| f \|_{2+\alpha}. \tag{3.24}$$

The estimations of other terms in (3.23) are contained in the following result.

Lemma 3.1. *One can choose* C > 0 *and* $\epsilon_o > 0$ *such that for any* $f \in H^{2+\alpha}$ *and* $\epsilon \leq \epsilon_o$,

$$\epsilon^2 \| G_m^\epsilon (C\Delta - L^{(N)}) f \|_{2+\alpha} \leq (1 - \frac{\gamma}{8C}) \| f \|_{2+\alpha} \tag{3.25}$$

and

$$\epsilon^2 \| G_m^\epsilon f \|_{2+\alpha} \leq \frac{\| f \|_{2+\alpha} \gamma^2}{2^{10} C^2 Z |m|^{2+\alpha}}. \tag{3.26}$$

Proof. Define the operator Π_K mapping a function $f(\varphi) = \sum_q f_q e^{i(q,\varphi)} \in H^\alpha$ to $\Pi_K f(\varphi) = \sum_{q:|q|\leq K} f_q e^{i(q,\varphi)}$, and let $\Pi^K f = f - \Pi_K f$. Then

$$\begin{aligned} G_m^\epsilon (C\Delta - L^{(N)}) f &= G_m^\epsilon \Pi_K (C\Delta - L^{(N)}) f \\ &+ G_m^\epsilon \Pi^K \Gamma^{(N)} \Pi^K f + G_m^\epsilon \Pi^K (C\Delta - L^{(N)}) \Pi_K f \\ &+ G_m^\epsilon \Pi^K (C\Delta - L^{(N)} - \Gamma^{(N)}) \Pi^K f, \end{aligned} \tag{3.27}$$

where $\Gamma^{(N)} = (f^{(N)}(\varphi), \nabla) = \sum_{k \leq \nu} b_N^k(\varphi) \frac{\partial}{\partial \varphi_k}$.

Put

$$\delta(R) = \min_{q: 0 < |q| \leq R} |(q,\omega)| \tag{3.28}$$

which is a positive nondecreasing in R function, provided the frequencies are rationally indpendent. By (3.17), (3.22), and (3.28) we have

$$\epsilon^2 \| G_m^{\epsilon} \, \Pi_K (C\Delta - L^{(N)}) f \|_{2+\alpha} \leq \frac{\epsilon^2 (C+Z)(K+N)^2}{\delta(K+|m|)} \| f \|_{2+\alpha}, \quad (3.29)$$

$$\epsilon^2 \| G_m^{\epsilon} \, \Pi^K (C\Delta - L^{(N)}) \, \Pi_K f \|_{2+\alpha} \leq \frac{\epsilon^2 (C+Z) K^2}{\delta(K+N+|m|)} \| f \|_{2+\alpha}, \quad (3.30)$$

and

$$\epsilon^2 \| G_m^{\epsilon} \, \Pi^K \Gamma^{(N)} \Pi^K \, f \|_{2+\alpha} \leq \frac{Z}{CK} \| f \|_{2+\alpha}. \quad (3.31)$$

If

$$h = \Pi^K (C\Delta - L^{(N)} - \Gamma^{(N)}) \, \Pi^K f \quad (3.32)$$

and the Fourier expansion of h has the form $h(\varphi) = \sum_q h_q e^{i(q,\varphi)}$ then

$$h_q = - C|q|^2 f_q + \sum_{p:\,|q-p|\leq N,\, p>K} f_p (A_{q-p} p, p)$$

if $|q| > K$ and $h_q = 0$, if $|q| \leq K$.
Set

$$V_{q,N} = \sum_{p:\,|q-p|\leq N,\, |p|>K} \bar{f}_p (\bar{A}_{q-p} p, p)$$

where $\bar{A}_q$ is a matrix with elements which are complex conjugates to the corresponding elements of A_q, then one can write

$$\|h\|_{\alpha}^2 = C^2 \| \Pi^K f \|_{2+\alpha}^2 + \sum_{q:\,|q|>K} |V_{q,N}|^2 |q|^{2\alpha} \quad (3.33)$$

$$- C \sum_{q:\,|q|>K} |q|^{2+2\alpha} (f_q V_{q,N} + \bar{f}_q \bar{V}_{q,N}).$$

Since $\overline{V}_{q,N}$ is the q-th Fourier coefficient of $\Pi^K L^{(N)} \Pi^K f$ then by the second inequality in (3.22) one has

$$\sum_{q:|q|>K} |V_{q,N}|^2 |q|^{2\alpha} = \| \Pi^K L^{(N)} f \|_\alpha^2 \le Z^2 \| \Pi^K f \|_{2+\alpha}^2 . \quad (3.34)$$

Furthermore,

$$W_{K,N} = \sum_{q:|q|\ge K} |q|^{2+2\alpha} f_q V_{q,N}$$

$$= \sum_{q:|q|\ge K} |q|^{2+2\alpha} f_q \sum_{p:|q-p|\le N, |p|>K} \overline{f}_p (\overline{A}_{q-p} p, p)$$

$$= W^{(1)} + W^{(2)} + W^{(3)}$$

where

$$W^{(1)} = \sum_{p,q:|q-p|\le N, |p|>K, |q|>K} f_q \overline{f}_p |q|^{1+\alpha} |p|^{1+\alpha} (\overline{A}_{q-p} q, p)$$

$$= \int_{\mathcal{T}^\nu} (\overline{A}^{(N)}(\varphi) v(\varphi), v(\varphi)) d\varphi$$

with $v(\varphi) = (v_1(\varphi), \cdots, v_\nu(\varphi))$ given by

$$v_\ell(\varphi) = \sum_{q:|q|>K} f_q |q|^{1+\alpha} q_\ell e^{i(q,\varphi)}, \quad q = (q_1, \cdots, q_\nu).$$

By $W^{(2)}$ and $W^{(3)}$ we have denoted the expressions

$$W^{(2)} = \sum_{p,q:|q-p|\le N, |p|>K, |q|>K} f_q \overline{f}_p |q|^{1+\alpha} (|q|^{1+\alpha} - |p|^{1+\alpha}) \times (\overline{A}_{q-p} p, p)$$

and

$$W^{(3)} = \sum_{p,q:\,|q-p|\leq N,\,|p|>K,\,|q|<K} f_q \overline{f}_p |q|^{1+\alpha} (\overline{A}_{q-p}(p-q),p).$$

By (3.21),

$$\mathrm{Re}W^{(1)} \geq \frac{\gamma}{2} \|\Pi^K f\|^2_{2+\alpha}. \tag{3.35}$$

If $|q-p|\leq N$, $0 \leq \alpha \leq 1$, and $|p| > 0$ then
$||q|^{1+\alpha} - |p|^{1+\alpha}| \leq ((\frac{N}{|p|} + 1)^{1+\alpha} - 1)\ |p|^{1+\alpha}$
$\leq ((\frac{N}{|p|} + 1)^2 - 1)|p|^{1+\alpha} \leq N(N+2)|p|^{\alpha}$.

Thus by the definition of Z,

$$|W^{(2)}|\leq ZN(N+2) \sum_{p,q:\,|q-p|\leq N,\,|p|>K,\,|q|>K} |f_q f_p| \times |q|^{1+\alpha}|p|^{2+\alpha}$$
$$\leq 2ZN(N+2)K^{-1} \sum_{p,q:\,|q-p|\leq N,\,|p|>K,\,|q|>K} (|f_q|^2|q|^{4+2\alpha}+|f_p|^2|p|^{4+2\alpha})$$
$$= 4ZN^{\nu+1}(N+2)K^{-1}\|\Pi^K f\|^2_{2+\alpha}. \tag{3.36}$$

Similarly,

$$|W^{(3)}| \leq ZN \times \sum_{p,q:\,|q-p|\leq N,\,|p|>K,\,|q|>K} |f_q f_p| \times |q|^{1+\alpha}|p|^{2+\alpha}$$
$$\leq 4ZN^{\nu+1}K^{-1}\|\Pi^K f\|^2_{2+\alpha}. \tag{3.37}$$

Since

$$\epsilon^2 \|G_m^\epsilon h\|_{2+\alpha} \leq C^{-1}\|h\|_\alpha$$

we derive from (3.32) - (3.37) that

$$\epsilon^2 \|G_m^\epsilon \Pi^K(C\Delta - L^{(N)} - \Gamma^{(N)})\Pi^K f\|_{2+\alpha} \tag{3.38}$$

$$\leq (1 + Z^2C^{-2} - \gamma(2C)^{-1} + 8ZN^{\nu+1}(N+2)(KC)^{-1})^{1/2}\|f\|_{2+\alpha}.$$

Put

$$K = [2^7 ZN^{\nu+1}(N+2)\gamma^{-1}] + 1 \text{ and } C = 2^4 Z^2 \gamma^{-1}, \tag{3.39}$$

where, again, $[\cdot]$ denotes the integral part.
Then

$$\epsilon^2 \|G_m^\epsilon \Pi^K(C\Delta - L^{(N)} - \Gamma^{(N)})\Pi^K f\|_{2+\alpha} \leq \tag{3.40}$$

$$\leq (1 - \frac{3\gamma}{8C})^{1/2} \|f\|_{2+\alpha} \leq (1 - \frac{3\gamma}{16C}) \|f\|_{2+\alpha}.$$

Set

$$\epsilon_1 = \left(\frac{\gamma \cdot \delta(K+N+|m|)}{64C(C+Z)(K+N)^2}\right)^{1/2}$$

then by (3.27) - (3.31), (3.39), and (3.40) we conclude that (3.25) holds true provided $0 < \epsilon \leq \epsilon_1$.

Next, by (3.17) and (3.28),

$$\epsilon^2 \|G_m^\epsilon f\|_{2+\alpha} \leq \epsilon^2 \|G_m^\epsilon \Pi_{\tilde{K}} f\|_{2+\alpha} + \epsilon^2 \|G_m^\epsilon \Pi^{\tilde{K}} f\|_{2+\alpha} \tag{3.41}$$

$$\leq \|f\|_{2+\alpha}(\epsilon^2(\delta(\tilde{K}+|m|))^{-1} + (C\tilde{K}^2)^{-1}),$$

and so (3.26) follows, provided $\tilde{K} = [(2^{11}CZ|m|^{2+\alpha}\gamma^{-2})^{1/2}] + 1$ and $0 < \epsilon \leq \epsilon_2 = \tilde{K}^{-1}(C^{-1}\delta(\tilde{K} + |m|))^{1/2}$.
Thus Lemma 3.1 is proved with C given by (3.39) and $\epsilon_0 = \min(\epsilon_1, \epsilon_2)$. □

Remark 3.1. The direct estimates in the proof of Lemma 3.1 can be replaced by an application of the general Garding inequality (see Yosida [Yos], Section 8 of Chapter 6) which leads to a shorter proof.

Next, we come back to the proof of Theorem 3.1. If $\epsilon \leq \epsilon_0$ then by (3.22) - (3.26),

$$\|\Phi_m^\epsilon f\|_{2+\alpha} \leq 1 + \left(1 - \frac{\gamma}{16C} + \frac{\gamma^2}{10^{10}C^2|m|^{2+\alpha}}\|f\|_{2+\alpha}\right)\|f\|_{2+\alpha} \quad (3.42)$$

It follows that for any m, $|m| \neq 0$,

$$\|\Phi_m^\epsilon f\|_{2+\alpha} \leq R_m \text{ provided } \|f\|_{2+\alpha} \leq R_m, \quad (3.43)$$

where $R_m = 32|m|^{2+\alpha}C\gamma^{-1}$, i.e., the operator Φ^ϵ_m maps into itself the ball S_m of the radius R_m centered at 0 in the space $H^{2+\alpha}$. If $|m| = 0$ then Theorem 3.1 is clear since $\lambda_0 = \lambda_0^\epsilon = 0$ and $r_0 = r_0^\epsilon = 1$ are the eigenvalue and the corresponding eigenfunction of both the operator B and the operator L^ϵ for all $\epsilon > 0$. Hence we must treat only the case $|m| \neq 0$.

We shall show next that Φ_m^ϵ is a contraction in the intersection of the ball S_m with the set of functions satisfying (3.16). Indeed, let f, q $\in S_m$ and $(f, r_m) = (g, r_m) = 1$, then by (3.22) - (3.26),

$$\|\Phi^\epsilon_m f - \Phi^\epsilon_m g\|_{2+\alpha} \le \epsilon^2\|G^\epsilon(CA - L^{(N)})(f-g)\|_{2+\alpha} \tag{3.44}$$

$$+ \epsilon^2\|G^\epsilon_m(L^{(N)}-L)(f-g)\|_{2+\alpha} + \epsilon^2|(L(f-g),r_m)|\|G^\epsilon_m f\|_{2+\alpha}$$

$$+ \epsilon^2|(Lg,r_m)|\|G^\epsilon_m(f-g)\|_{2+\alpha}$$

$$\le (1 - \frac{3\gamma}{32C} + \frac{2R_m\gamma^2}{2^{10}C^2|m|^{2+\alpha}})\|f-g\|_{2+\alpha} = (1 - \frac{\gamma}{32C})\|f-g\|_{2+\alpha}$$

Hence Φ^ϵ_m is a contraction and since Φ^ϵ_m satisfies (3.16) then it follows that Φ^ϵ_m has in $S_m \cap \{f\in H^{2+\alpha}: \ (f,r_m) = 1\}$ the unique fixed point r^ϵ_m, i.e., (3.14) holds true. Thus (3.12) and (3.13) are also satisfied, and since $\|r^\epsilon_m\|_{2+\alpha} \le R_m$ we derive

$$|\lambda^\epsilon_m - i(m,\omega)| \le \epsilon^2\|Lr^\epsilon_m\|_\alpha \le \epsilon^2 ZR_m. \tag{3.45}$$

It is not difficult to see that, in fact, the operator Φ^ϵ_m has a single fixed point in a ball of radius $R_m(\epsilon)$ where $R_m(\epsilon) \longrightarrow \infty$ as $\epsilon \longrightarrow 0$.

Since $\Phi^\epsilon_m r^\epsilon_m = r^\epsilon_m$ then by (3.17) and (3.18) similarly to (3.41) one has

$$\|r^\epsilon_m - r_m\|_{2+\alpha-\beta} \le \epsilon^2 \|\Pi_{K(\epsilon)}G^\epsilon_m(CA-L)r^\epsilon_m\|_{2+\alpha-\beta} \tag{3.46}$$

$$+ \epsilon^2(Lr^\epsilon_m,r_m)\|\Pi_{K(\epsilon)}G^\epsilon_m r^\epsilon_m\|_{2+\alpha-\beta}$$

$$+ \epsilon^2\|\Pi^{K(\epsilon)}G^\epsilon_m(CA-L)r^\epsilon_m\|_{2+\alpha-\beta}$$

$$+ \epsilon^2(Lr^\epsilon_m,r_m)\|\Pi^{K(\epsilon)}G^\epsilon_m r^\epsilon_m\|_{2+\alpha-\beta}$$

$$\le \epsilon^2((K(\epsilon))^{2-\beta}(C+Z)R_m + ZR^2_m)(\delta(K(\epsilon) + |m|))^{-1}$$

$$+ ((1 + ZC^{-1})R_m + ZC^{-1}R^2_m)(K(\epsilon))^{-2\beta}.$$

We choose $K(\epsilon)$ so that

$$\delta(K(\epsilon) + |m|) \geq \epsilon^{\frac{1}{2}}, \quad (K(\epsilon))^{2-\beta}((C+Z)R_m + ZR_m^2) \leq \epsilon^{-\frac{1}{2}} \tag{3.47}$$

and $K(\epsilon) \longrightarrow \infty$ as $\epsilon \longrightarrow 0$. The speed of growth of $K(\epsilon)$ as $\epsilon \longrightarrow 0$ depends on the speed of decrease of $\delta(R)$ as $R \longrightarrow \infty$, i.e., it depends on the approximative properties of the frequenceis $(\omega_1, \cdots, \omega_\upsilon)$. This completes the proof of the assertion (i) of Theorem 3.1.

To obtain the assertion (ii) suppose that an eigenvalue λ^ϵ of L^ϵ has the form $\lambda^\epsilon = \alpha^\epsilon + i\ \beta^\epsilon$ where α^ϵ and β^ϵ are real numbers and $0 \geq \alpha^\epsilon \geq -C^{(1)}\ \epsilon^2$, $C^{(1)} \geq 0$. We introduce the operator $\mathcal{U}^\epsilon$: $H^{2+\alpha} \longrightarrow H^{2+\alpha}$ acting by the formula $\mathcal{U}^\epsilon f = g$, where

$$\epsilon^2 C\Delta g + Bg - i\beta^\epsilon g = \epsilon^2(C\Delta - L)\ f + \alpha^\epsilon f. \tag{3.48}$$

Choose now C,N, and K in the same way as in the proof of (i), take $\tilde{K} = [(64C^{(1)}\gamma^{-1})^{1/2}] + 1$, and put $Q = \max\ (K + N, \tilde{K}.)$ Suppose that,

$$\min_{q:|q|\leq Q} |\beta^\epsilon - (q,\omega)| \geq C^{(2)}\epsilon^2 \tag{3.49}$$

where $C^{(2)} > 0$ is a big number given by (3.50) below. Then the inequalities (3.29) and (3.30) will remain true with G_m^ϵ replaced by $G^\epsilon = (\epsilon^2 C\Delta + B - i\beta^\epsilon)^{-1}$ and $\delta(K+|m|)$ or $\delta(K+N+|m|)$ replaced by $C^{(2)}\epsilon^2$. Thus if

$$C^{(2)} = 64C(C+Z)(K+N)^2(C^{(1)} + 1)\gamma^{-1} \tag{3.50}$$

then in the same way as in the proof of Lemma 3.1 we shall obtain (3.25) with G_m^ϵ replaced by G^ϵ defined above.

Next, in the same way as in (3.41),

$$\|G^{\epsilon}(\alpha^{\epsilon}f)\|_{2+\alpha} \le C^{(1)}\epsilon^{2}(\|G^{\epsilon}\Pi_{\tilde{K}}f\|_{2+\alpha} + \|G^{\epsilon}\Pi^{\tilde{K}}f\|_{2+\alpha})$$

$$\le C^{(1)}((C^{(2)})^{-1} + (C\tilde{K}^{2})^{-1})\|f\|_{2+\alpha}$$

which in view of (3.50) and the definition of $\tilde{K}$ leads to the inequality

$$\|G^{\epsilon}(\alpha^{\epsilon}f)\|_{2+\alpha} \le \frac{\gamma}{32C}\|f\|_{2+\alpha}. \qquad (3.51)$$

This together with (3.24) and (3.25) where G^{ϵ}_{m} is replaced by G^{ϵ} give

$$\|\mathfrak{A}^{\epsilon}f\|_{2+\alpha} \le (1 - \frac{\gamma}{32C})\|f\|_{2+\alpha}. \qquad (3.52)$$

It follows from (3.52) that under the condition (3.49) there is no solution of the quation $\mathfrak{A}^{\epsilon}f^{\epsilon} = f^{\epsilon}$ except zero. Thus for some q with $|q| \le Q$ the inequality $|\beta^{\epsilon} - (q,\omega)| \le C^{(2)}\epsilon^{2}$ must hold which together with (3.45) proves the assertion (ii) and completes the proof of Theorem 3.1. □

Remark 3.2. If the frequencies $(\omega_1,\cdots,\omega_\nu)$ are in resonance the situation differs from the one considered in Theorem 3.1. Firstly, infinitely many eigenvalues of the operator L^{ϵ} may converge to one eigenvalue of the operator $B = (\omega_1 \frac{\partial}{\partial\varphi_1},\cdots,\omega_\nu \frac{\partial}{\partial\varphi_\nu})$ which can be seen in the case of an operator L having constant coefficients. Secondly, (3.8) may fail, as one can see in the case of the operator

$$L^{\epsilon} = \epsilon^{2}((2 + e^{-i\varphi_2})\frac{\partial^{2}}{\partial\varphi_1^{2}} + (2 + e^{-i\varphi_1})\frac{\partial^{2}}{\partial\varphi_2^{2}}) + \frac{\partial}{\partial\varphi_1}$$

where all eigenfunction f^{ϵ} satisfying $(f^{\epsilon}, e^{i(m\varphi_1+n\varphi_2)}) = 1$ tend to the infinity in the norm of H^2 with the speed equal to const.$\times\ \epsilon^{-1}$ (see [Ki4], Remark 2).

Next, we shall discuss pertubations of operators with continuous spectra. We shall consider the discrete time case. Let F be a homeomorphism of a compact space M preserving a probability measure μ on M. The operator $\mathcal{U}$ acting by the formula

$$\mathcal{U}f(x) = f(F^{-1}x) \tag{3.53}$$

is unitary in the Hilbert space H with the inner product

$$(f,g) = \int_M f(x)\ \overline{g(x)}d\mu(x).$$

The spectrum of the dynamical system $\{F^n,\ n = 0,\pm1,\pm2,\cdots\}$ is defined as the spectrum of the operator $\mathcal{U}$ (see Cornfeld, Fomin, and Sinai [CFS], part III).

Let Q be a completely continuous self adjoint operator in H having a positive continuous kernel q(x,y) so that

$$Qf(x) = \int_M q(x,y)f(y)d\mu(y) \text{ and } Q1 = 1. \tag{3.54}$$

According to the Hilbert-Schmidt theorem (see Yosida [Yos], Chapter 11) the operator Q has a complete orthonormal system of real eigenfunctions $f_0 = 1, f_1, f_2, \cdots$ and corresponding eigenvalues $e^{\lambda_0}, e^{\lambda_1}, \cdots$ with $\lambda_0 = 0$, $\lambda_k < 0$ for $k \neq 0$, and $\lambda_k \longrightarrow -\infty$ as $k \longrightarrow \infty$. It is easy to see that the kernel q(x,y) can be represented in the form

$$q(x,y) = \sum_{k=0}^{\infty} e^{\lambda_k} f_k(x)f_k(y). \tag{3.55}$$

Define an operator P by the formula $Pf(x) = Qf(Fx)$ and consider a Markov chain X_n with transition probabilities $P(\xi,\Gamma) = P\chi_\Gamma(x)$ where, again, χ_Γ is the indicator function of a set Γ. We may view X_n as a random perturbation of the homeomorphism F.

The spectrum of the operator P can be studied in the following way. Let $g \in H$ and $g(x) = \sum_{k=0}^{\infty} g_k f_k(x)$ then by (3.55),

$$Pg(x) = \int_M q(Fx,y)\, g(y) d\mu(y) = \sum_{k=0}^{\infty} e^{\lambda_k} f_k(Fx) g_k. \quad (3.56)$$

Put

$$\psi_{mn} = e^{\lambda_n} \int_M f_m(x) f_n(Fx) d\mu(x)$$

and let Ψ be the infinite matrix with elements ψ_{mn}. If

$$Pg(x) = \alpha g(x) \quad (3.57)$$

then, in view of (3.55), the equality (3.56) is equivalent to the relation

$$\Psi \hat{g} = \alpha \hat{g} \quad (3.58)$$

where $\hat{g} = (g_0, g_1, \cdots)$ is the vector whose components are the coefficients of the expansion $g(x) = \Sigma\, g_k f_k(x)$. Usually, it is not easy to determine the eigenvalues of the matrix Ψ, but under certain conditions this can be done.

Theorem 3.2. *Suppose that the operator $\mathcal{U}$ defined by (3.53) has a continuous spectrum on the subspace orthogonal to constants, i.e., the family of projectors $R(\theta)$ in the spectral representation*

$$\mathcal{U} = \int_0^{2\pi} e^{i\theta} dR(\theta)$$

of the operator $\mathcal{U}$ *(see Yosida [Yos], Chapter 11) depends continuously on* θ. *Assume that for all* $k = 0,1,\cdots$

$$f_k(Fx) = \gamma_k \, f_{\varphi(k)}(x) \text{ for some } \gamma_k, |\gamma_k| = 1 \qquad (3.59)$$

where φ: $\{0,1,2,\cdots\} \longrightarrow \{0,1,2,\cdots\}$, $\varphi(0) = 0$ *is a one-to-one correspondence. Then the spectrum of the operator* P *on the subspace orthogonal to constants consists of the single point* 0.

Proof. The operator P is completely continuous, and so (see Yosida[Yos], Chapter 10) it has pure point spectrum apart perhaps from 0. Let α be an eigenvalue of P, i.e.,

$Pg = \alpha g$ for some $g(x) = \sum_{k=1}^{\infty} g_k f_k(x)$ with $(g,g) = 1$. By (3.56) and (3.59) for any integers $k,\ell > 0$ we obtain

$$\alpha^{\ell} g_k = g_{\varphi^{-\ell}(k)} \exp\{\sum_{j=1}^{\ell} \lambda_{\varphi^{-j}(k)}\} \prod_{j=1}^{\ell} \gamma_{\varphi^{-j}(k)} \qquad (3.60)$$

Since $\mathcal{U}^n$ has the spectral representation

$$\mathcal{U}^n = \int_0^{2\pi} e^{in\theta} dR(\theta) \quad , \; n = \pm 1, \pm 2, \cdots,$$

(see Yosida [Yos], Chapter 11) then our assumption that $\mathcal{U}$ has a continuous spectrum is equivalent to the requirement that $\mathcal{U}$ and all its powers have no eigenfunctions (except for contants). Thus by (3.59) for any $k > 0$,

$$\varphi^{-\ell}(k) \longrightarrow \infty \text{ as } \ell \longrightarrow \infty. \qquad (3.61)$$

Indeed, if (3.61) were not true then we would have a subsequence ℓ_j, $j = 1,2,\cdots$ such that for some $k > 0$ the sequence $\varphi^{-\ell_j}(k)$ stays in a bounded set. Hence for some $n_1 > n_2$,

$\varphi^{-n_1}(k) = \varphi^{-n_2}(k)$, i.e., $k = \varphi^{n_1-n_2}(k)$. But then it follows from (3.53) and (3.59) that $f_k(x)$ is an eigenfunction of the operator $\mathcal{U}^{n_1-n_2}$ which contradicts our assumption and proves (3.61).

Since the eigenvalues e^{λ_j} of Q tend to zero as $j \longrightarrow \infty$, $|\gamma_j| = 1$ and $|g_j| \leq 1$ for all $j = 0,1,2,\cdots$ then (3.60) and (3.61) imply that $\alpha = 0$. From (3.60) it is clear also that the number 0 is not an eigenvalue of the operator P. In order to decide whether 0 belongs to the spectrum or not one has to study the equation $Pg = h$ which is equivalent to the system

$$h_k = g_{\varphi^{-1}(k)} \exp\{\lambda_{\varphi^{-1}(k)}\} \gamma_{\varphi^{-1}(k)}, \quad k = 0,1,2,\cdots \qquad (3.62)$$

where

$$h(x) = \sum_{k=0}^{\infty} h_k f_k(x) \text{ and } g(x) = \sum_{k=0}^{\infty} g_k f_k(x).$$

By (3.62),

$$|g_k| = h_{\varphi(k)} e^{\lambda_k}$$

and, since $\lambda_k \longrightarrow -\infty$ as $k \longrightarrow \infty$, there exists $h \in H$ for which the function g satisfying (3.62) does not belong to H. Therefore 0 belongs to the spectrum of $\mathcal{U}$ (in fact, it is easy to see that 0 belongs to the residual spectrum of $\mathcal{U}$), completing the proof of Theorem 3.2. □

The natural field of applications of Theorem 3.2 is the case of affine transformations of compact Abelian groups with the function $q(x,y)$ invariant under shifts. Let, for instance, F be an automorphism of the ν-dimensional torus $M = \mathcal{T}^{\nu}$ given by a matrix with integer elements and the determinant equal to one. Suppose that $q(x,y) = q(x-y)$.

In the case when the eigenvalues of F differ from 1 in absolute value (hyperbolic case), and also in the case when there are eigenvalues of each type, i.e., which have absolute values greater than 1, less than 1, and equal to 1 but which are not roots of 1 (partially hyperbolic case) the corresponding unitary operator is known to have a continuous spectrum, and so all conditions of Theorem 3.2 will be satisfied. Theorem 3.2 can be applied also in the case of transformations with quasi-discrete spectra, provided we consider them on a subspace of functions orthogonal to eigenfunctions.

The behavior of the spectrum exhibited in Theorem 3.2 seems to be rather special. It is not clear what usually happens when $\epsilon \longrightarrow 0$ with the spectra of transition operators P_t^{ϵ} of random perturbations X_t^{ϵ} of flows F^t. If X_t^{ϵ} is a diffusion process on a compact manifold M with a generator $L^{\epsilon} = \epsilon^2 L + B$ where B satisfies (1.1) then the problem is about asymptotical behavior as $\epsilon \longrightarrow 0$ of the spectrum of L^{ϵ}. When F^t has a pure-point (discrete) spectrum this problem was addressed in Theorem 3.1. On the other hand, if F^t has a continuous spectrum the problem is far from any satisfactory solution. Remark that when L^{ϵ} is considered on a compact manifold then the principal eigenvalue is zero (since $L^{\epsilon}1 = 0$). The real part of the eigenvalue with the second largest real part give a bound for the speed of convergence as $t \longrightarrow \infty$ of transition probabilities $P^{\epsilon}(t,x,\cdot)$ to the invariant measure of X_t^{ϵ}. The probabilistic sense of other eigenvalues is not clear.

Chapter IV

Random Perturbations of Some Special Models

In this chapter we shall study the convergence of invariant measures for random perturbations of maps of an interval and a Lorentz's type model dynamical system. These models lack the shadowing property for some pseudo-orbits. Misiurewicz's map treated in Section 4.2 is also not uniformly expanding. However, we shall see how to modify the approach of Chapter II in order to overcome these complications.

4.1. Random perturbations of one-dimensional transformations.

In this section we shall discuss models of random perturbations of maps of an interval. We shall treat specifically the case of the Lasota-Yorke [LaY] piecewise smooth expanding transformations.

Suppose that $F:[0,1]\to[0,1]$ is a map of the unit interval $M=[0,1]$ into itself. Random perturbations X_n^ϵ of F are being defined in the same way as in Section 1.1 though we shall impose certain restrictions on measures Q_x^ϵ and transition probabilities $P^\epsilon(x,\cdot)$ appearing in (I.1.1) and (I.1.2). We shall be interested in the weak convergence of invariant measures μ^ϵ of Markov chains X_n^ϵ (i.e., the probability measures on M satisfying (I.1.3)) as $\epsilon\to 0$.

In the same way as in Assumption II.1.1 (a) we shall assume that all Q_x^ϵ have densities q_x^ϵ with respect to the Lebesgue measure on $M = [0,1]$, i.e., for any Borel set $\Gamma \subset M$,

$$Q_x^\epsilon (\Gamma) = \int_\Gamma q_x^\epsilon (y)\, dy, \tag{1.1}$$

and so $P^\epsilon(x,\cdot) = Q_{Fx}^\epsilon (\cdot)$ has the density

$$p^\epsilon(x,y) = q_{Fx}^\epsilon (y). \tag{1.2}$$

Before we proceed with our further assumptions let us discuss first another model due to Boyarsky [Boy] and considered later also by Golosov [Gol] and Collet [Col]. They assume that $q_x^\epsilon(y) = q^\epsilon(y-x)$, i.e., this density depends only on the difference $(y-x)$. In view of (I.1.3), (1.1), and (1.2) any invariant measure μ^ϵ of X_n^ϵ has a density ρ^ϵ with respect to the Lebesgue measure which satisfies the equation

$$\rho^\epsilon(z) = \int_M \rho^\epsilon(x)\, p^\epsilon(x,z)\, dx, \tag{1.3}$$

and so if $q_x^\epsilon (y) = q^\epsilon (y-x)$ then

$$\rho^\epsilon(z) = \int_M \rho^\epsilon(x)\, q^\epsilon(z - Fx)\, dx \tag{1.4}$$

$$= \int_M \left[\sum_{x \in F^{-1}y} \rho^\epsilon(x)\, |F'(x)|^{-1} \right] q^\epsilon(z - y)\, dy$$

provided the derivative $F'(x)$ exists for almost all with respect to the Lebesgue measure points $x \in M = [0,1]$. Introduce the operator Φ acting on integrable functions g on $[0,1]$ by the formula

$$\Phi g(y) = \sum_{x \in F^{-1}y} (g(x)|F'(x)|^{-1}) \qquad (1.5)$$

which is called the Frobenius-Perron operator of the map F. This operator describes the transformation of the density of an absolutely continuous measure under the action of F, i.e., if $g = \frac{d\nu}{dx}$, $\nu \in \mathcal{P}(M)$ then Φg is the density of the measure ζ such that $\zeta(\Gamma) = \nu(F^{-1}\Gamma)$ for any Borel set $\Gamma \subset M$. Thus if there exists an absolutely continuous F-invariant measure then the density of this measure must be a fixed point of the operator Φ. Vice versa any fixed point of Φ which is an integrable function turns out to be the density of an absolutely continuous invariant measure. By this reason the study of the Frobenius-Perron operator plays a decisive part in many works concerning absolutely continuous measures of one-dimensional transformations (see Lasota and Yorke [LaY], Misiurewicz [Mi], Collet and Eckmann [CE1] and [CE2]).

Define another operator $\mathcal{P}^{\epsilon}$ acting on integrable functions by the formula

$$\mathcal{P}^{\epsilon} g(z) = \int_M (\Phi g(y))\, q^{\epsilon}(z-y)\, dy \qquad (1.6)$$

which may be called the Frobenius-Perron operator of random perturbations. In view of (1.4) and (1.5) the density ρ^{ϵ} is a fixed point of the operator $\mathcal{P}^{\epsilon}$ whose explicit representation (1.6) in a convolution form enables one to obtain uniform in ϵ estimates of its fixed points essentially in the same way as one estimates variations of fixed points of the operator Φ itself. By this technique Boyarsky [Boy], Golosov [Gol], and Collet [Col] showed for certain types of maps F that limits of invariant measures μ^{ϵ} of X_n^{ϵ} as $\epsilon \to 0$ must be absolutely continuous.

Moreover, by this method one can show the convergence of the densities ρ^{ϵ}, as well. However, the condition $q_x^{\epsilon}(y) = q^{\epsilon}(y-x)$ is rather restrictive and, as we shall see, it excludes interesting models where X_n^{ϵ} is obtained by means of a composition of maps chosen independently at random from a parametric family. Besides, the above approach cannot be generalized to arbitrary manifolds.

Next, we shall specify our conditions which are a one-dimensional version of Assumption II.1.1.

Assumption 1.1 (a) Transition probabilities of Markov chains X_n^{ϵ} have the form $P^{\epsilon}(z,\cdot) = Q_{Fz}^{\epsilon}(\cdot)$ with Q_x^{ϵ} satisfying (1.1);

(b) There exist constants α, $C > 0$, $\alpha < 1$ and a family of non-negative functions $\{r_x(\xi),\ x\in M = [0,1],\ \xi\in\mathbb{R}^1 = (-\infty,\infty)\}$ such that

$$q_x^{\epsilon}(y) \le C\epsilon^{-1} e^{-\frac{\alpha}{\epsilon}\mathrm{dist}(x,y)} \quad \text{for all} \quad x,y\in M, \tag{1.7}$$

where

$$\mathrm{dist}(x,y) = \min(|y-x|,\ |y-x+1|,\ |y-x-1|), \tag{1.8}$$

and

$$q_x^{\epsilon}(y) \le (1+\epsilon^{\alpha})\epsilon^{-1}\ r_x\!\left(\frac{1}{\epsilon}\,\sigma(x,y)\right) \tag{1.9}$$

provided $\mathrm{dist}(x,y) \le \epsilon^{1-\alpha}$, where $\sigma(x,y)$ equals one of the numbers $(y-x)$, $(y-x+1)$, or $(y-x-1)$ so that $|\sigma(x,y)| = \mathrm{dist}(x,y)$;

(c) The functions $r_x(\xi)$, $x\in M$, $\xi\in\mathbb{R}^1$ satisfy

(i) $\displaystyle\int_{\mathbb{R}^1} r_x(\xi)\, d\xi = 1$,

(ii) $r_x(\xi) \le C\, e^{-\alpha|\xi|}$ for α, $C>0$

independent of x and ξ.

(iii) There exists $C>0$ such that if

$V_x^+ = \{\xi : r_x(\xi) > 0\}$ and $\partial V_x^+(\delta)$ denotes the δ-neighborhood in $\mathbb{R}^1$ of the boundary ∂V_x^+ of V_x^+ then

$$\int_{\partial V_x^+(\delta)} r_x(\xi)\, d\xi \leq C\delta, \tag{1.10}$$

and

$$r_x(\xi) \leq r_y(\zeta) + C\rho + \chi_{\partial V_x^+(C\rho)}(\xi) r_x(\xi) \tag{1.11}$$

where $\rho = \rho((x,\xi),(y,\xi)) = \mathrm{dist}(x,y) + |\zeta-\xi|$.

Remark 1.1. The definition (1.8) of the distance means that we consider the periodic boundary conditions, i.e., that we identify the endpoints 0 and 1. Another boundary condition which can be treated by our method is the reflection condition in the endpoints 0 and 1. This means that (1.9) remains the same for either $x \in [\epsilon^{1-\alpha}, 1-\epsilon^{1-\alpha}]$ or $x < \epsilon^{1-\alpha}$ and $y \geq x$ or $x > 1-\epsilon^{1-\alpha}$ and $y \leq x$. But if $x < \epsilon^{1-\alpha}$ and $y < x$ then one assumes

$$\epsilon q_x^\epsilon(y)\left(r_x\left[\frac{y-x}{\epsilon}\right] + r_x\left[-\frac{(x+y)}{\epsilon}\right]\right)^{-1} \leq 1 + \epsilon^\alpha, \tag{1.12}$$

and if $x > 1-\epsilon^{1-\alpha}$ and $y > x$ then

$$\epsilon q_x^\epsilon(y)\left(r_x\left[\frac{y-x}{\epsilon}\right] + r_x\left[\frac{2-(x+y)}{\epsilon}\right]\right)^{-1} \leq 1 + \epsilon^\alpha \ . \tag{1.13}$$

In this case (1.7) should be replaced by

$$q_x^\epsilon(y) \leq C\epsilon^{-1} \exp\left(-\frac{\alpha}{\epsilon}|y-x|\right) \tag{1.14}$$

if $|y-x| > \epsilon^{1-\alpha}$ and $\epsilon > 0$ is small enough. We can treat also the situation when $\operatorname{dist}(x,y) = |x-y|$ and $q_x^\epsilon(y)$ equals zero unless $x \in [0,1]$ and y belongs to an open neighborhood U of $[0,1]$. This must be complemented by the condition $FU \subset [0,1]$ which yields Markov chains X_n^ϵ defined on U. Boundary conditions do not influence decisively the study of corresponding random perturbations and related proofs differ only in details.

Remark 1.2. For a justification of the condition (iii) we refer the reader to Remark II.1.1. Note that Examples II.1.1-II.1.3 satisfy our conditions. Since we identify the endpoints, and so M turns out to be a circle, we have to take in Example II.1.3 a diffusion on this circle.

In this section we shall work with transformations satisfying the following conditions.

Assumption 1.2. A map F is piecewise C^2 and expanding, i.e., there exist points $0 = \alpha_0 < \alpha_1 < \cdots < \alpha_{\upsilon+1} = 1$ such that the restrictions of F to the open intervals (a_{i-1}, a_i), $i = 1, \cdots, \upsilon + 1$, $\upsilon \geq 1$ are C^2 functions which can be extended to the closed intervals $[a_{i-1}, a_i]$ as C^2 functions (taking at the endpoints right or left derivatives), and

$$\inf_x |F'(x)| = \lambda > 1 \tag{1.15}$$

where the infinum is taken over all $x \in [0,1]$ for which the derivative $F'(x)$ exists.

Under Assumption 1.2 F is known to have invariant measures which are absolutely continuous with respect to the Lebesgue measure on [0,1] (see Lasota and York [LaY] and Cornfeld, Fomin, and Sinai [CFS], § 4 of Chapter 7). Li and York [LiY] showed that in the above situation there exist at most υ ergodic absolutely continuous F-invariant probability measures. In particular if $\upsilon = 1$ then one has only one absolutely continuous F-invariant probability measure.

In order to avoid certain complications we shall assume that F is continuous with respect to the metric defined by (1.8), in particular, $F(0) = F(1)$. We shall establish the following result.

Theorem 1.1. *Suppose that random perturbations* X_n^ε *of a map* $F:[0,1]\ [0,1]$ *meet the conditions of Assumption 1.1,* F *is continuous with respect to the dist-metric, and* F *satisfies Assumption 1.2. Then all weak limits as* $\varepsilon\to 0$ *of probability invariant measures* μ^ε *of Markov chains* X_n^ε *are absolutely continuous with respect to the Lebesgue measure* [0,1]. *In particular, if* $\nu = 1$ *in Assumption 1.2 then the invariant measures* μ^ε *weakly converge as* $\varepsilon\to 0$ *to the unique absolutely continuous F-invariant probability measure* μ.

Before the proof we shall discuss certain points connected with Theorem 1.1 for the one-parameter family of tent maps

$$F_s x = \begin{cases} sx & \text{if } 0 \le x \le \frac{1}{2} \\ s(1-x) & \text{if } \frac{1}{2} \le x \le 1 \end{cases} \tag{1.16}$$

where $1 < s \le 2$. These maps meet conditions of Assumption 1.2 with $\nu = 1$ and $a_1 = \frac{1}{2}$. First, notice that in general these maps do not have the shadowing property. Indeed, let $F = F_{\sqrt{2}}$. Then $c = 2-\sqrt{2}$ is a repelling fixed point of F and $F^3(\frac{1}{2}) = c$. Take the δ-pseudo-orbit $x_o = \frac{1}{2}$, $x_1 = F_x$, $x_2 = F_{x_1}$, $x_3 = c$, $x_4 = c+\delta$, $x_5 = F(x_4),\cdots$, $x_{k+1} = Fx_k,\cdots$, $k = 4,5,\cdots$. Consider the interval $I = \{x: |x - \frac{1}{2}| \le \frac{1}{4}(3-2\sqrt{2})\}$ then $F^3(I)$ is the interval whose left endpoint is c, and so $F^4(I)$ is the interval whose right endpoint is c. Hence if $y \in I$ then $x_4 > c \ge F^4 y$. Since $|x_{k+1} - F^{k+1}y| = \sqrt{2}|x_k - F^k y|$ for $k \ge 4$ provided $\frac{1}{2} \ge x_k \ge \frac{\sqrt{2}}{2}$ and $\frac{1}{2} \ge F^k y \ge \frac{\sqrt{2}}{2}$ we conclude that the orbit of y cannot shadow in any reasonable sense the δ-pseudo-orbit $x_0, x_1, x_2, \cdots$ when δ is small enough. Therefore we shall need some substitution for the shadowing property when proving Theorem 1.1.

For the family of tent maps F_s, $1 < s \leq 2$ we can consider the following model of random perturbations. Suppose that $1 < s_0 < 2$ and let $\varphi_1^\epsilon, \varphi_2^\epsilon, \cdots$ be independent random variables with the same uniform distribution on the interval $[-\epsilon, \epsilon]$ where $0 < \epsilon \leq \epsilon_0 < \min\left[s_0 - 1, \dfrac{s_0(2-s_0)}{(s_0+2)}\right]$. Consider the Markov chains

$$X_n^\epsilon = F_{s_0+\varphi_n^\epsilon} \circ \cdots \circ F_{s_0+\varphi_1^\epsilon} x \quad . \tag{1.17}$$

Let $a_\epsilon = s_0(1 - \frac{1}{2}(s_0 + \epsilon)) - \epsilon$ and $b_\epsilon = \frac{1}{2}(s_0 + \epsilon)$ then $0 < a_\epsilon < \frac{1}{2} < b_\epsilon < 1$. Moreover if $x \in [a_\epsilon, b_\epsilon]$ then $X_n^\epsilon \in [a_\epsilon, b_\epsilon]$ for all $n = 1, 2, \cdots$. Thus we can study the invariant measures of X_n^ϵ on the interval $[a_\epsilon, b_\epsilon]$. Notice that transition probabilities of Markov chains X_n^ϵ have the form

$$P^\epsilon(x, \Gamma) = \int_\Gamma q_{Fx}^\epsilon(y)\, dy$$

where

$$q_z^\epsilon(y) = \begin{cases} s_0(2\epsilon z)^{-1} & \text{if } |y-z| \leq \epsilon z s_0^{-1} \\ 0 & \text{if } |y-z| > \epsilon z s_0^{-1} \end{cases} \tag{1.18}$$

Thus this type of random perturbations considered on $[a_{\epsilon_0}, b_{\epsilon_0}]$ with $\epsilon < \epsilon_0$ satisfy conditions of Assumption 1.1 and an application of Theorem 1.1 yields that invariant measures μ^ϵ of X_n^ϵ on $[a_{\epsilon_0}, b_{\epsilon_0}]$ weakly converge as $\epsilon \to 0$ to the unique F_{s_0}-invariant absolutely continuous probability measure. Remark, that $q_z^\epsilon(y)$ given by (1.18) depends essentially on z, and so it cannot be represented as $q^\epsilon(y-z)$. Hence the Boyarsky's approach based on the Frobenius-Perron operator and described at the beginning of this section does not work in this situation. If $s_o = 2$ then the above type of random perturbations formally does not satisfy Assumption 1.1 since in this case we cannot exclude the point 0 which is the fixed point for all F_s, $1 < s < 2$, and so $P^\epsilon(0,\cdot)$ will not have density at all. Moreover, the unit mass at 0 is the invariant measure for any X_n^ϵ, $\epsilon > 0$, and so Theorem 1.1 is not true, as stated, on the whole interval [0,1]. However, if $s_o = 2$ and $\varphi_1^\epsilon, \varphi_2^\epsilon, \cdots$ are independent random variables uniformly distributed on $[-\epsilon, 0]$ with $\epsilon > 0$ small enough, then a careful study by our method of approaches of X_n^ϵ close to 0 yields that invariant measures μ^ϵ of X_n^ϵ having no atom at 0 weakly converge as $\epsilon\to 0$ to the Lebesque measure on [0,1] which is F_2-invariant.

Next, we shall prove a version of the shadowing property which will be sufficient for our purposes.

Lemma 1.1. *Suppose that* $x_o, x_1, x_2, \cdots, x_n$ *is a* δ*-pseudo-orbit, i.e.,* (I.4.1) *holds true with* dist *defined by* (1.8) *and* δ *small enough. If*

$$\min_{0\le i\le n-1}\ \min_{0\le j\le \nu+1} \operatorname{dist}(x_i, a_j) \ge \delta(\lambda-1)^{-1} \tag{1.19}$$

with $a_o, \cdots, a_{\nu+1}$ *introduced in Assumption* 1.2 *and* λ *from* (1.15), *then there exists a point* $y \in [0,1]$ *such that*

$$\operatorname{dist}(F^i y, x_i) \le \delta(\lambda-1)^{-1} \text{ for all } i = 0,\cdots,n. \tag{1.20}$$

Proof. Notice that the failure of shadowing which we exhibited in the case of tent maps is due, in fact, to the existence of points in [0,1] having no preimages. But under our conditions a δ-pseudo-orbit may contain such points only if it contains also points close to a_j, $j = 0, \cdots, \upsilon+1$ which we prohibit. By this reason if (1.19) holds true then for any $k = 0, 1, \cdots, n-1$ the interval $J_k = \{z: \text{dist}(z, x_k) \leq \delta(\lambda-1)^{-1}\}$ contains a connected component of the set $F^{-1}J_{k+1}$ provided δ is small enough. Then the intersection $\bigcap_{k=0}^{n} F^{-k}J_k$ is not empty and any point belonging to this intersection can play the role of y in (1.20). □

Next, we proceed as follows.

Lemma 1.2. *There exists* $C > 0$ *such that for any* $\gamma > 0$, $\gamma < 1$, *each* $x \in [0,1]$, *and an interval* $Q \subset [0,1]$ *one has*

$$I_o^\epsilon(\epsilon^{1-\gamma}, n, x, Q) \leq C \text{ mes } Q \tag{1.21}$$

provided $(\log \epsilon)^4 \geq n \geq (\log \epsilon)^2$ *and* ϵ *is small enough, where* mes *is the Lebesque measure on* $[0,1]$ *and*

$I_o^\epsilon(\rho, n, x, \Gamma) = P_x^\epsilon\{\min_{0 \leq k \leq n-1} \min_{0 \leq j \leq \upsilon+1} \text{dist}(X_k^\epsilon, a_j) \geq \rho$ *and* $X_n^\epsilon \in \Gamma\}$.

Proof. We employ the same arguments as in Section 2.4. for the case of expanding transformations. Put

$$I_1^\epsilon(\rho, \delta, n, x, \Gamma) = P_x^\epsilon\{\text{dist}(FX_k^\epsilon, X_{k+1}^\epsilon) < \delta \text{ and } \tag{1.22}$$

$$\min_{0 \leq k \leq n-1} \min_{0 \leq j \leq \upsilon+1} \text{dist}(X_k^\epsilon, a_j) \geq \rho \text{ and } X_n^\epsilon \in \Gamma\}.$$

Then similarly to Lemma II.1.1 it follows that

$$|I_o^\epsilon(\epsilon^{1-\gamma}, n, x, Q) - I_1^\epsilon(\epsilon^{1-\gamma}, \epsilon^{1-\beta}, n, x, Q)| \tag{1.23}$$

$$\leq (\text{mes } Q) \exp(-\alpha/3\epsilon^\beta)$$

provided ϵ is small enough.

If $\operatorname{mes} Q \geq \epsilon$ then one can choose points $v_1, \cdots, v_{\ell_\epsilon}$ such that

$$Q \subset \bigcup_{1 \leq i \leq \ell_\epsilon} U_\epsilon(v_i) \text{ and } \operatorname{mes} Q \geq \frac{1}{2} \sum_{1 \leq i \leq \ell_\epsilon} \operatorname{mes} U_\epsilon(v_i) \qquad (1.24)$$

where $U_\rho(v) = \{w: \operatorname{dist}(w,v)<\rho\}$.
If $\operatorname{mes} Q<\epsilon$ then we put $\ell_\epsilon = 1$ and take Q itself in place of $U_\epsilon(v_1)$. Thus

$$I_1^\epsilon(\epsilon^{1-\gamma}, \epsilon^{1-\beta}, n, x, Q) \leq \sum_{1 \leq i \leq \ell_\epsilon} I_1^\epsilon(\epsilon^{1-\gamma}, \epsilon^{1-\beta}, n, x, U_\epsilon(v_i)). \qquad (1.25)$$

The probability $I_1^\epsilon(\epsilon^{1-\gamma}, \epsilon^{1-\beta}, n, x, U_\epsilon(v_i))$ involves only $\epsilon^{1-\beta}$-pseudo-orbits $\omega = (x_0, \cdots, x_n)$, $x_k = X_k^\epsilon$, which do not approach the points $a_j, j = 0, \cdots, \nu+1$ closer than $\epsilon^{1-\gamma}$, and so if $\gamma > \beta$, say $\gamma = 3\beta$, and ϵ is small enough then we can employ Lemma 1.1 to find a point $y = y^\omega$ whose orbit shadows this $\epsilon^{1-\beta}$-pseudo-orbit in the sense of (1.20) with $\delta = \epsilon^{1-\beta}$. We proceed in the same way as in the proof of Theorem II.4.1 by choosing points z_{ijk}, $j \leq \epsilon^{-2\beta}+1$ such that one point z_{ijk} is taken in each connected component of the intersection $(\overline{U_{j\epsilon}(x) \setminus U_{(j-1)\epsilon}(x)}) \cap F^{-n} U_\epsilon(v_i)$. Since $n \geq (\log \epsilon)^2$ then we conclude similarly to (II.4.14) that

$$I_1^\epsilon(\epsilon^{1-3\beta}, \epsilon^{1-\beta}, n, x, U_\epsilon(v_i)) \leq \qquad (1.26)$$

$$\sum_{j \leq \epsilon^{-2\beta}, k} I_2^\epsilon(\epsilon^{1-2\beta}, n, x, z_{ijk}, U_\epsilon(v_i))$$

where the sum is over z_{ijk} such that the orbit $F^\ell z_{ijk}, \ell=0, \cdots, n$ does not approach the points $a_j, j=0, \cdots, \nu+1$ closer than $\epsilon^{1-2\beta}$, and

$$I_2^\epsilon(\rho,n,x,z,\Gamma)=P_x^\epsilon\{\text{dist}(X_\ell^\epsilon,F^\ell z)\le\rho \text{ for all } \ell=0,\cdots,n \quad (1.27)$$

$$\text{and } X_n^\epsilon\in\Gamma\} = \int_{U_\rho(Fz)}\cdots\int_{U_\rho(F^{n-1}z)}\int_{U_\rho(F^n z)\cap\Gamma}$$

$$q_{Fx}^\epsilon(y_1)\cdots q_{Fy_{n-1}}^\epsilon(y_n)dy_1\cdots dy_n.$$

Let $2\beta<\alpha$ then by (1.9) taking into account that the orbit of $z=z_{ijk}$ stays $\epsilon^{1-2\beta}$-apart from 0 and 1 we obtain that

$$I_2^\epsilon(\epsilon^{1-2\beta},n,x,z,U_\epsilon(v_i))\le(1+\epsilon^\alpha)^n\int_{U_{\epsilon^{1-2\beta}}(Fz)} \quad (1.28)$$

$$\cdots\int_{U_{\epsilon^{1-2\beta}}(F^{n-1}z)}\int_{U_{\epsilon^{1-2\beta}}(F^n z)\cap U_\epsilon(v_i)}\epsilon^{-1}r_{Fx}\left[\frac{y_1-Fx}{\epsilon}\right]$$

$$\epsilon^{-1}r_{Fy_1}\left[\frac{y_2-Fy_1}{\epsilon}\right]\cdots\epsilon^{-1}r_{Fy_{n-1}}\left[\frac{y_n-Fy_{n-1}}{\epsilon}\right]dy_1\cdots dy_n.$$

Since F has a bounded second derivative apart from the points $a_j,j=0,\cdots,\nu+1$ and intervals $(y_\ell,F^\ell z)$ do not contain these points then

$$|y_{\ell+1}-Fy_\ell-(y_{\ell+1}-F^{\ell+1}z)+F'(F^\ell z)(y_\ell-F^\ell z)|\le\epsilon^{2-5\beta} \quad (1.29)$$

provided ϵ is small enough and $\ell=0,\cdots,n-1$. Next, we set $\eta_\ell=y_\ell-F^\ell z$ and replace in the right hand side of (1.28) each $r_{Fy_\ell}\left[\frac{y_{\ell+1}-Fy_\ell}{\epsilon}\right]$ by $r_{F^{\ell+1}z}\left[\frac{(\eta_{\ell+1}-F'(F^\ell z)\eta_\ell}{\epsilon}\right]$ which acccording to (1.11) and (1.29) may decrease the right hand side of (1.28) by no more than a positive power of ε provided $0 < \beta < \frac{1}{5}$ (cf.(II.4.25)). In view of (1.15) we can employ the one-dimensional counterparts of Propositions II.2.1 and II.3.11 to derive (1.21) from (1.28) and the above arguments precisely in the same way as in the proof of Theorem II.4.1 using obvious simplifications due to the one-dimensional situation. □

To complete the proof of Theorem 1.1 we shall need the following result which enables us to estimate the probabilities of arriving to small neighborhoods of the points $a_j, j=0,\cdots,\nu+1$ which we had to exclude in Lemma 1.2.

Lemma 1.3. *There exists* $C > 0$ *such that if* $\gamma > 0$ *is small enough then for any* $x,y \in [0,1]$ *and* $k \geq \log(\frac{1}{\varepsilon})$

$$P^{\varepsilon}(k,x,U_{\varepsilon^{1-\gamma}}(y)) \leq C\varepsilon^{\gamma} \quad . \tag{1.30}$$

Proof. By the Chapman-Kolmogorov formula for any $\ell < k$ one has

$$P^{\varepsilon}(k,x,U_{\varepsilon^{1-\gamma}}(y)) = \int_{[0,1]} P^{\varepsilon}(k-\ell,x,dz)P^{\varepsilon}(\ell,z,U_{\varepsilon^{1-\gamma}}(y)) \tag{1.31}$$

$$\leq \sup_{z\in[0,1]} P^{\varepsilon}(\ell,z,U_{\varepsilon^{1-\gamma}}(y)) \quad ,$$

and so if (1.30) will be proved for $k = \ell$ then it will remain true for any $k \geq \ell$.

Without loss of generality we can assume that $\lambda > 2$ in (1.15). Indeed, we can always choose an integer $r > 0$ such that $\lambda^r > 2$ and then pass to Markov chains $Y_n^{\varepsilon} = X_{nr}^{\varepsilon}$ which are random perturbations of the map F^r satisfying Assumption 1.1 since F is Lipschitz continuous. The assertion of Lemma 1.3 proved for Y_n^{ε} will imply the desired assertion for X_k^{ε} itself. Thus we assume that $\lambda > 2$.

Under Assumption 1.2 we have

$$\sup_{x} |F'(x)| = D < \infty \quad . \tag{1.32}$$

Put

$$\ell = [\tfrac{1}{2}\beta(\log(D + 1))^{-1} \log(\tfrac{1}{\epsilon})] \quad , \tag{1.33}$$

where $1 > \beta > 0$ and $[\cdot]$ means the integral part. Then

$$(D + 1)^{\ell+1} \leq \epsilon^{-\beta/2} \quad . \tag{1.34}$$

In the same way as in the previous lemma we can pass to $\epsilon^{1-\beta}$-pseudo-orbits making only a negligible mistake in our estimates. In view of (1.32), (1.34), and the continuity of F every $\epsilon^{1-\beta}$-pseudo-orbit $y_o = x, y_1, \cdots, y_\ell$ satisfies

$$\operatorname{dist}(F^i x, y_i) < \epsilon^{1-2\beta} \text{ for all } i=0,1,\cdots,\ell \quad . \tag{1.35}$$

We shall introduce points $z_i, i=0,\cdots,\ell$ by setting $z_i = F^i x$ if $\min_{0 \leq j \leq \nu+1} \operatorname{dist}(a_j, F^i x) \geq \epsilon^{1-2\beta}$ and $z_i = a_j$ if $\operatorname{dist}(a_j, F^i x) < \epsilon^{1-2\beta}$. For ϵ small enough these define points z_i uniquely. By (1.35),

$$\max_{0 \leq i \leq \ell} \operatorname{dist}(z_i, y_i) < 2\epsilon^{1-2\beta} \tag{1.36}$$

but for $z_i = F^i x$ we have a better inequality (1.35). Thus taking into account (1.7), (1.9), and the Chapman-Kolmogorov formula we can write

$$P^\epsilon(\ell, x, \Gamma) \leq I_3^\epsilon(\ell, x, \Gamma) + \exp(-\alpha/3\epsilon^\beta) \tag{1.37}$$

where $2\beta < \alpha$,

$$I_3^\epsilon(\ell,x,\Gamma) = P_x^\epsilon\{X_i^\epsilon \in U^{(i)} \text{ for all } i=0,1,\cdots,\ell \text{ and } X_\ell^\epsilon \in \Gamma\} \quad (1.38)$$

$$= \int_{U^{(0)}} \cdots \int_{U^{(\ell-1)}} \int_{U^{(\ell)}\cap\Gamma} q_{Fx}^\epsilon(y_1) q_{Fy_1}^\epsilon(y_2) \cdots q_{Fy_{\ell-1}}^\epsilon(y_\ell)$$

$$dy_1 \cdots dy_\ell \le (1+\epsilon^\alpha)^\ell \int_{U^{(0)}} \cdots \int_{U^{(\ell-1)}} \int_{U^{(\ell)}\cap\Gamma} \epsilon^{-1} r_{Fx}\left[\frac{\sigma(Fx,y_1)}{\epsilon}\right]$$

$$\times\, \epsilon^{-1} r_{Fy_1}\left[\frac{\sigma(Fy_1,y_2)}{\epsilon}\right] \cdots \epsilon^{-1} r_{Fy_{\ell-1}}\left[\frac{\sigma(Fy_{\ell-1},y_\ell)}{\epsilon}\right] dy_1 \cdots dy_\ell \ .$$

where $U^{(i)} = \{v: \ \mathrm{dist}(v,z_i) < \epsilon^{1-2\beta}\}$ if $z_i = F^i x$ and $U^{(i)} = \{v: \ \mathrm{dist}(v,z_i) < 2\epsilon^{1-2\beta}\}$ if $z_i \ne F^i x$.

We cannot proceed precisely in the same way as in the proof of Lemma 1.2 using (1.29) since the derivative of F may have discontinuities at the points $a_j, j=0,\cdots,\nu+1$.

Let $F'_+(z_i)$ and $F'_-(z_i)$ be the right and the left derivatives of F at z_i, respectively. Put $\eta_i = y_i - z_i$, $b_i = z_{i+1} - Fz_i$, $A(\eta_i) = F'_+(z_i)$ if $\eta_i \ge 0$ and $A(\eta_i) = F'_-(z_i)$ if $\eta_i < 0$, $i = 0,\cdots,\ell$. Then for ϵ small enough and $y_i \in U^{(i)}$, $i = 1,\cdots,\ell$ one has

$$|\sigma(Fy_i,y_{i+1}) - \eta_{i+1} + A(\eta_i)\eta_i - b_i| \le \epsilon^{2-5\beta}, \quad (1.39)$$

$i = 0,\cdots,\ell-1$ since by our choice of the points z_i the map F can be extended as a C^2 function into each closed interval $[y_i,z_i]$.

Replace in the right hand side of (1.38) each

$$r_{Fy_i}\left[\frac{\sigma(Fy_i,y_{i+1})}{\epsilon}\right] \text{ by } r_{z_{i+1}}\left[\frac{\eta_{i+1} - A(\eta_i)\eta_i + b_i}{\epsilon}\right] .$$

According to (1.11) and (1.39) this substitution may decrease the right hand side of (1.38) by no more than a positive power of ϵ provided $\beta < \frac{1}{5}$. These will lead to an expression which can be bounded by the integral

$$I_4^\epsilon(\ell,x,\Gamma) = \int_{\mathbb{R}^1} \cdots \int_{\mathbb{R}^1} \int_{\Gamma} \epsilon^{-1}\, r_{z_1}\left[\frac{\eta_1+b_0}{\epsilon}\right]$$

$$\times\, \epsilon^{-1}\, r_{z_2}\left[\frac{\eta_2-A(\eta_1)\eta_1+b_1}{\epsilon}\right] \qquad (1.40)$$

$$\times\cdots\times\, \epsilon^{-1} r_{z_\ell}\left[\frac{\eta_\ell-A(\eta_{\ell-1})\eta_{\ell-1}+b_{\ell-1}}{\epsilon}\right] d\eta_1\cdots d\eta_\ell .$$

Define inductively

$$\Xi_{k+1}^\epsilon = A(\Xi_k^\epsilon)\, \Xi_k^\epsilon + \epsilon\theta_{k+1} - b_k \qquad (1.41)$$

where $\Xi_o^\epsilon = 0$ and $\theta_1, \theta_2, \cdots$ are independent random variables with the distributions

$$P\{\theta_k \in \Psi\} = \int_{\Psi} r_{z_k}(\eta)d\eta \quad . \qquad (1.42)$$

It is easy to see in the same way as in Lemma II.2.1 that Ξ_k^ϵ $k = 0,1,\cdots$ is a nonhomogeneous Markov chain whose transition density from $\Xi_k^\epsilon = \eta$ to $\Xi_{k+1}^\epsilon = \eta$ has the form $\epsilon^{-1}\, r_{z_{k+1}}\left[\frac{\zeta-A(\eta)\eta+b_k}{\epsilon}\right]$. Thus

$$I_4^\epsilon(\ell,x,\Gamma) = P\{\Xi_\ell^\epsilon \in \Gamma\} \quad . \qquad (1.43)$$

By (1.41),

$$\Xi_\ell^\epsilon = \epsilon \sum_{k=1}^{\ell-1} A(\Xi_{\ell-1}^\epsilon)\cdots A(\Xi_k^\epsilon)(\theta_k - \epsilon^{-1}b_{k-1}) + \epsilon\theta_\ell - b_{\ell-1}. \quad (1.44)$$

For any sequence $\kappa = (\kappa_1, \cdots, \kappa_{\ell-1})$ $\kappa_i = \pm 1$ of $+1$ and -1 we put $B_i(\kappa) = F'_+(z_i)$ if $\kappa_i = +1$ and $B_i(\kappa) = F'_-(z_i)$ if $\kappa_i = -1$. Denote also

$$\Xi_\ell^\epsilon(\kappa) = \epsilon \sum_{k=1}^{\ell-1} B_{\ell-1}(\kappa)\cdots B_k(\kappa)(\theta_k - \epsilon^{-1} b_{k-1}) + \epsilon\theta_\ell - b_{\ell-1}. \quad (1.45)$$

Since by (1.15), $|B_i(\kappa)| \geq \lambda$ for all i and κ we derive in the same way as in the proof of (II.2.7) that

$$P\{\Xi_\ell^\epsilon(\kappa) \in \Gamma\} \leq \tilde{C}\epsilon^{-1}\lambda^{-\ell} \text{ mes } \Gamma \quad (1.46)$$

for all $\kappa = (\kappa_1, \cdots, \kappa_{\ell-1})$. Then we obtain

$$P\{\Xi_\ell^\epsilon \in \Gamma\} \leq \sum_\kappa P\{\Xi_\ell^\epsilon(\kappa) \in \Gamma\} \leq \tilde{C}\epsilon^{-1}\left[\frac{2}{\lambda}\right]^\ell \text{ mes } \Gamma . \quad (1.47)$$

Since we assume that $\lambda > 2$ then taking $\Gamma = U_{\epsilon^{1-\gamma}}(y)$ and ℓ from (1.33) we derive (1.30) from (1.31), (1.37), and the arguments following after (1.37) together with (1.43) and (1.47), provided $\gamma > 0$ is small enough. □

Now we can complete the proof of Theorem 1.1. From Lemma 1.3 and the Markov property it follows that

$$P_x^\epsilon\left\{\min_{n \geq k \geq \log\left[\frac{1}{\epsilon}\right]} \min_{\upsilon+1 \geq j \geq 0} \text{dist}(X_k^\epsilon, a_j) < \epsilon^{1-\gamma}\right\} \leq Cn\epsilon^\gamma . \quad (1.48)$$

Thus taking $n = n(\epsilon) = [2(\log \epsilon)^2]$ and $\ell = \ell(\epsilon) = [\log \frac{1}{\epsilon}] + 1$ we obtain from (1.21), (1.48), and the Markov property that for any interval $Q \subset [0,1]$,

$$P^\epsilon(n,x,Q) \le P^\epsilon_x\{X^\epsilon_n \in Q \text{ and } \min_{n\ge k\ge \log(\frac{1}{\epsilon})} \min_{\nu+1\ge j\ge 0} \tag{1.49}$$

$$\operatorname{dist}(X^\epsilon_k, a_j) \ge \epsilon^{1-\gamma}\} + \epsilon^{\gamma/2} = \int_{[0,1]} P^\epsilon(\ell,x,dy)$$

$$I^\epsilon_o(\epsilon^{1-\gamma}, n-\ell, y, Q) + \epsilon^{\gamma/2} \le C \text{ mes } Q + \epsilon^{\gamma/2}$$

provided ϵ is small enough. Hence for any invariant probability measure μ^ϵ of X^ϵ_n one has

$$\mu^\epsilon(Q) = \int_{[0,1]} d\mu^\epsilon(x)\, P^\epsilon(n,x,Q) \le C \text{ mes } Q + \epsilon^{\gamma/2}\,. \tag{1.50}$$

which being true for any interval $Q \subset [0,1]$ implies the assertion of Theorem 1.1. □

Adapting the arguments of Section 2.6 to the one-dimensional piecewise C^2 and expanding maps F in the same way as above one obtains the entropy convergence result provided $\nu = 1$.

Theorem 1.2. *Let* $\Pi = (Q_1,\cdots,Q_k)$ *be a partition of* $[0,1]$ *into intervals with sufficiently small lengths and suppose that* $\nu = 1$ *which means that invariant measures* μ^ϵ *of* X^ϵ_n *weakly converge to the unique absolutely continuous F-invariant measure* μ. *Then*

$$\lim_{\epsilon\to 0} h^\epsilon(\theta,\zeta^\epsilon) = h_\mu(F) \tag{1.51}$$

where ζ^ϵ *is the partition of the sample space* Ω *into the sets* $\Gamma^\epsilon_j = \{\omega\colon\ X^\epsilon_o \in Q_j\}$, $h_\mu(F)$ *is the entropy of* F *relative to the measure* μ, *and* $h^\epsilon(\theta,\zeta^\epsilon)$ *is the entropy of the shift transformation* θ *defined in Section 1.2.*

4.2. Misiurewicz's maps of an interval.

In this section we shall discuss certain points concerning random perturbations of one-dimensional maps satisfying Misiurewicz's conditions from [Mi]. For the detailed exposition we refer the reader to Katok and Kifer [KK].

We shall consider random perturbations X_n^ϵ satisfying Assumption 1.1 of maps F of the interval having a non-positive Schwarzian derivative, no sinks and future orbits of critical points staying away from critical points. We shall restrict ourselfs to the most widely considered one- parameter family of maps

$$F_\lambda : \quad x \to 4\lambda x(1-x) \tag{2.1}$$

for which the above conditions are satisfied for a set of parameters λ having cardinality of the continuum.

Assumption 2.1. The map F_λ has the form (2.1), it has no stable periodic orbit, and

$$\frac{1}{2} \notin \mathcal{T}_\lambda = \overline{\bigcup_{n\geq 1} F_\lambda^n \left(\frac{1}{2}\right)} \quad . \tag{2.2}$$

According to Misiurewicz [Mi] any map F_λ satisfying Assumption 2.1 possess exactly one absolutely continuous invariant measure μ_{F_λ} which is ergodic.

The following result was proved in Katok and Kifer [KK].

Theorem 2.1. *Suppose that random perturbations* X_n^ϵ *of the map* F_λ *meet the conditions of Assumption 1.1 and* F_λ *satisfies Assumption 2.1. Then invariant measures* μ^ϵ *of* X_n^ϵ *weakly converge to* μ_{F_λ} *as* $\epsilon \to 0$.

Example 2.1. Similarly to the previous section we can consider the following model of random perturbations satisfying our conditions. Let $\varphi_1, \varphi_2, \cdots$ be independent random variables with the same distribution having a smooth density $\rho(x)$ concentrated on $[-1,1]$. Suppose that λ_o is a fixed parameter such that $\frac{1}{2} < \lambda_o < 1$ and the map F_{λ_o} satisfies the above conditions. Then for $\varepsilon < 1-\lambda_o$ the composition of independent random transformations $F_{\lambda_o+\varepsilon\varphi_i}$, $i=1,2,\cdots$ generates a Markov chain $X_n^{\varepsilon} = F_{\lambda_o+\varepsilon\varphi_n} \circ \cdots \circ F_{\lambda_o+\varepsilon\varphi_1} x$ which, considered on the invariant interval $[4\lambda_o(\lambda_o+\varepsilon)(1-\lambda_o-\varepsilon)-\varepsilon,\ \lambda_o+\varepsilon]$, belongs to the class of random perturbations satisfying our conditions. The case $\lambda_o = 1$ must be studied separately since then we cannot exclude from the consideration the point 0 which is fixed for all F_λ. The transition probability of X_n^{ε} can be written in the form

$$P^{\varepsilon}(x,\Gamma) = P\{F_{\lambda_o+\varepsilon\varphi_i} x\in\Gamma\} = P\{\varepsilon\varphi_i \in (x(1-x))^{-1}\Gamma - \lambda_o\} \quad (2.3)$$

$$= (4\varepsilon x(1-x))^{-1} \int_{\Gamma} \rho\left[\frac{1}{\varepsilon}\left[\frac{y}{4x(1-x)} - \lambda_o\right]\right] dy \quad .$$

Thus the transition density $p^{\varepsilon}(x,y) = (4\varepsilon x(1-x))^{-1}$ $\rho\left[\frac{1}{\varepsilon}\left[\frac{y-F_{\lambda_o}x}{4x(1-x)}\right]\right]$ does not have the form $q^{\varepsilon}(y-F_{\lambda_o}x)$ needed for an application of the Frobenius-Perron operator method described at the beginning of the previous section.

Remark 2.1. The stability of measures μ_{F_λ} with respect to random perturbations is especially interesting in view of the fact that in general there is no stability with respect to deterministic perturbations in this case. Indeed, consider F_λ with λ close to 1. Clearly, F_1 satisfies Assumption 2.1 and it has absolutely continuous invariant measure with the density $\pi^{-1}(x(1-x))^{-1/2}$. Put

$n_\lambda = \min\{n>1: F_\lambda^n(\frac{1}{2}) \geq \frac{1}{2}\}$. Since $F_1^n(\frac{1}{2}) = 0$ for all $n>1$ then if $F_\lambda^{n_\lambda}(\frac{1}{2}) > \frac{1}{2}$ by the continuity one can find $\beta(\lambda)$ such that $1 > \beta(\lambda)>\lambda$ and $F_{\beta(\lambda)}^{n_\lambda}(\frac{1}{2}) = \frac{1}{2}$. Hence $\frac{1}{2}$ is a periodic point of $F_{\beta(\lambda)}$ and its orbit is attracting since $F_\lambda'(\frac{1}{2}) = 0$ for any λ. Thus we obtained a sequence $\lambda_k \uparrow 1$ such that any F_{λ_k} has an attracting periodic orbit containing $\frac{1}{2}$ and only one point of this orbit can be to the right of $\frac{1}{2}$. The invariant measure ν_{λ_k} supported by this periodic orbit is stable with respect to random perturbations since the complement of its basin of attraction has zero Lebesque measure (see Collet and Eckmann [CE1], Proposition II.5.7). On the other hand, measures ν_k do not converge as $\lambda_k \to 1$ to the smooth invariant measure of F_1 since the above periodic orbits have only one point to the right of $\frac{1}{2}$ and so all weak limits of ν_k have support in the interval $[0,\frac{1}{2}]$. Similar examples can be constructed for $\lambda_k \to \lambda_o \neq 1$ with λ_o satisfying Assumption 2.1.

The maps F_λ do not necessarily have the shadowing property for all pseudo-orbits. However one can obtain the following result (see Katok and Kifer [KK], Lemma 2.3).

Lemma 2.1. *Suppose that* F_λ *satisfies Assumption 2.1 and let* $x_o, \cdots, x_n$ *be a* ε^β*-pseudo-orbit of* F_λ, *i.e.*, *(I.1.4) holds true with* $F = F_\lambda$ *and dist defined by (1.8). There exists a constant* $C > 0$ *depending only on* F_λ *such that if* $0 \leq \gamma \leq \beta/2$ *and*

$$|x_k - \frac{1}{2}| \geq 2C\varepsilon^\gamma, \quad k = 0,\cdots,n-1 \qquad (2.4)$$

then one can find a point y *so that*

$$\mathrm{dist}(F_\lambda^k y, x_k) \leq C\varepsilon^{\beta-\gamma}, \quad k = 0,\cdots,n. \qquad (2.5)$$

Since $F'_\lambda(\frac{1}{2}) = 0$ then, of course, the maps F_λ are not expanding. However, Assumption 2.1 yields some substitution for expanding which turns out to be sufficient both for Lemma 2.1 and other aspects of our approach.

Lemma 2.2. *Suppose that* F_λ *satisfies Assumption 2.1. There exists* $\eta > 1$ *such that for any* $\rho > 0$ *one can find an integer* $M_\rho > 0$ *so that*

$$|(F_\lambda^{M_\rho})'(x)| \geq \eta \quad \textit{provided} \quad \min_{0 \leq i < M_\rho} |F_\lambda^i x - \tfrac{1}{2}| \geq \rho \qquad (2.6)$$

and

$$|(F_\lambda^n)'(x)| \quad \textit{provided} \quad \mathrm{dist}(F_\lambda^n x, \mathcal{G}_\lambda) \geq \rho \qquad (2.7)$$

for any $x \in [0,1]$ *and* $n \geq 1$.

For the proof we refer the reader to Misiurewicz [Mi], Theorem 1.3 and to Katok and Kifer [KK], Lemma 2.2.

Under Assumption 2.1 the map F_λ becomes expanding in the sense that $|(F_\lambda^n)'(x)|$ grows exponentially fast in n for points x whose orbit stay away from $\mathcal{G}_\lambda$. Indeed, suppose that $F_\lambda^k x \neq \frac{1}{2}$ for all $k = 0,1,\cdots,n$. While $F_n^k x$ is not too close to $\frac{1}{2}$ then the derivative grows exponentially fast by (2.6). If for some k, $|F_\lambda^k x - \frac{1}{2}| = \rho$ then

$$|(F_\lambda^{k+1})'x| = 8\lambda\rho|(F_\lambda^k)'x| \qquad (2.8)$$

and

$$\mathrm{dist}\,(F_\lambda^{k+1}\,x,\ \mathcal{G}_\lambda) \leq \mathrm{dist}(F_\lambda^{k+1}\,x,\ F_\lambda(\tfrac{1}{2})) = 4\lambda\rho^2\ . \qquad (2.9)$$

Thus in view of (2.2) and (2.9) in order to have another chance to get close to $\frac{1}{2}$ the orbit must accumulate the derivative of order ρ^{-2} which according to (2.6) will take of order $\log(\frac{1}{\rho})$ steps. If $\ell_\rho = C_1 \log(\frac{1}{\rho})$ is this number of steps then $|(F^{k+\ell_\rho+1})'x| = |(F_\lambda^k)'x|C_2\rho^{-1}$ $= |(F_\lambda^k)'x|\, C_2(e^{1/c_1})^{\ell_\rho}$ which again leads to the exponential growth.

Still, proceeding with our method one has to face certain complications due to small derivatives of F_λ near $\frac{1}{2}$. Lemma 2.1 enables us to employ the linearization procedure if we restrict ourselfs to paths of X_n^ϵ which are ϵ^β-pseudo-orbits staying outside of the $2C\epsilon^\gamma$-neighborhood of the point $\frac{1}{2}$. However, this will lead to orbits of F_λ which may approach $\frac{1}{2}$ as close as $C(2\epsilon^\gamma+\epsilon^{\beta-\gamma})$, and so the derivatives of F_λ^k may be sometimes that small. By this reason a direct counterpart of Proposition II.2.1 will not work here. The following result proved in Appendix to Katok and Kifer [KK] saves the situation.

Lemma 2.3. *Suppose in addition to Assumption 1.1 that for each* $x \in [0,1]$ *the number of points of discontinuity of* $r_x(\xi)$ *in* ξ *is bounded by a number* N *independent of* x *and on each interval of continuity* $r_x(\xi)$ *is Lipschitz continuous in* ξ. *For arbitrary points* $x_1,\cdots,x_n \in [0,1]$ *let* $\theta_1,\cdots,\theta_n$ *be independent random variables with distribution functions* $P\{\theta_i \le a\} = \int_{-\infty}^{a} r_{x_i}(\xi)d\xi$. *Then there exist* $C,\kappa > 0$ *independent of* $x_1,\cdots,x_n$ *and* n *such that for any nonzero numbers* $a_1,\cdots,a_n$ *the distribution function of the random variable*

$$\left[\sum_{1\le i\le n} a_i^2\right]^{-1/2} \sum_{1\le i\le n} a_i(\theta_i - E\theta_i)$$

has the derivative, i.e., the probability density function, satisfying

$$r_{x_1,\cdots,x_n}^{a_1,\cdots,a_n}(\zeta) \le Ce^{-\kappa|\zeta|}$$

where $E\theta_i$ *is the expectation of* θ_i.

We discussed here only few arguments involved in the proof of Theorem 2.1 which is pretty long and can be found in Katok and Kifer [KK].

Remark 2.2. One can adapt the arguments of Section 2.6 and prove Theorem 1.2 of the previous section also for maps F_λ satisfying Assumption 2.1.

Remark 2.3. Another class of maps with a critical point (and so not uniformly expanding) possessing absolutely continuous invariant measures was studied by Collet and Eckmann [CE2]. For instance, for a set of parameters having a positive Lebesque measure the one-parameter family of maps F_λ: $[0,1] \to [0,1]$ given by the formula

$$F_\lambda x = \begin{cases} 1-2|x-\frac{1}{2}| & \text{if} \quad |x-\frac{1}{2}| \geq \lambda \\ 1-\lambda-(x-\frac{1}{2})^2 \lambda^{-1} & \text{if} \quad |x-\frac{1}{2}| \leq \lambda, \end{cases}$$

$0 < \lambda < \frac{1}{2}$ satisfies the conditions of [CE2]. Collet [Col] studied random perturbations of Boyarsky's type for this class of maps employing the Frobenius-Perron operator method described at the beginning of Section 1.1. It is not difficult to adapt the machinery of Katok and Kifer [KK] in order to prove Theorem 2.1 for this class of maps employing results of Appendices A and E from Collet [Col] which actually provide necessary dynamical prerequisites for our approach similar to Section 2 of Katok and Kifer [KK].

4.3. Lorenz's type models.

In this section we shall discuss random perturbations of model dynamical systems which are believed to describe main features of the Lorenz attractor (see Guckenheimer and Holmes [GH] or Sparrow [Sp]).

In 1963 E. Lorenz [Lo] published a paper describing a qualitative study by numerical integration of the following three-dimensional system of ordinary differential equations with three parameters $\sigma, r, b > 0$,

$$\frac{dx}{dt} = \sigma(y-x)$$

$$\frac{dy}{dt} = rx - y - xz \qquad (3.1)$$

$$\frac{dz}{dt} = xy - bz$$

derived from a model of fluid convection. Computer experiments indicated that for certain choice of parameters $\sigma, r,$ and b the flow F^t generated by (4.1) has an attractor (called now Lorenz's) where orbits of F^t exhibit a chaotic behavior.

The divergence of the vector field $(\sigma(y-x), rx-y-xz, xy-bz)$ equals $-(\sigma+1+b)$, and so F^t contracts the volume by $e^{-(\sigma+1+b)t}$ for $t>0$. Furthermore, consider the Lyapunov function $V(x,y,z) = rx^2+\sigma y^2+\sigma(z-2r)^2$ then

$$\left.\frac{dV(F^t(x,y,z))}{dt}\right|_{t=0} = -2\sigma(rx^2+y^2+bz^2-2brz) . \qquad (3.2)$$

Let c be the maximum of V in the bounded domain where $\frac{dV}{dt} \geq 0$. If $\delta > 0$ is small enough then it is easy to see that all orbits of F^t eventually enter the bounded ellipsoid $\mathcal{E} = \{(x,y,z): V(x,y,z) \leq c + \delta\}$. Thus we conclude that all orbits tend towards a bounded set of zero volume (see Sparrow [Sp], Appendix C).

Let X_t^ε be diffusion random perturbations of the flow F^t described in Example II.1.3. For any $\rho>0$ we can consider Markov chains $Y_n^{\varepsilon,\rho} = X_{n\rho}^\varepsilon$ which are random perturbations of $F = F^\rho$. In view of (3.2) it is easy to see that conditions of Theorem I.1.7 are satisfied for Markov chains $Y_n^{\varepsilon,\rho}$. Thus all their invariant measures have support in $\mathcal{E}$ and when $\epsilon \to 0$ then all weak limits of these measures are supported by a bounded set of zero volume. In particular, this is true for invariant measures μ^{ϵ} of diffusion processes X_t^{ϵ}.

The most popular choice of parameters leading to what is usually called the Lorenz attractor is $\sigma = 10$, $r = 28$ and $b = 8/3$. The origin 0 is the stationary point of the hyperbolic type for the system (3.1). It has the two-dimensional stable manifold $W^s(0)$ and the one-dimensional unstable manifold consisting of two branches Γ_1 and Γ_2. The plane $\Pi = \{(x,y,z): z = 27\}$ contains two more hyperbolic fixed points 0_1 and 0_2 which have one-dimensional stable manifolds which are lines contained in Π and two-dimensional unstable manifolds transverse to Π. The following picture illustrates the situation.

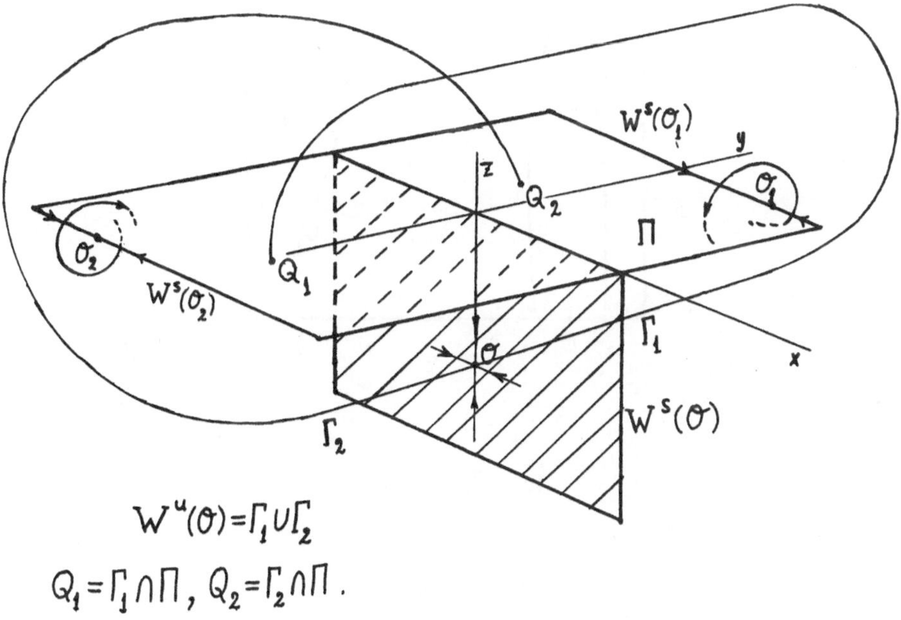

Figure 3.1.

Next, one considers the Poincare return map G of the plane Π to itself. Namely, if v is a point on this plane and the integral curve containing v goes downwards when intersecting the plane Π at v then Gv is the point of the next intersection of the integral curve with Π. The map G is not defined on the intersection $W^s(0) \cap \Pi$ and G maps points approaching this intersection from one side close to $Q_1 = \Gamma_1 \cap \Pi$ while points approaching $W^s(0) \cap \Pi$ from another side are being mapped close top1393Xbeing $Q_2 = \Gamma_2 \cap \Pi$.

By a change of coordinates we can reduce the study to the transformation G mapping the square $S = \{(x,y): |x| \leq 1, |y| \leq 1\} \subset \Pi$ into itself as shown on the following picture.

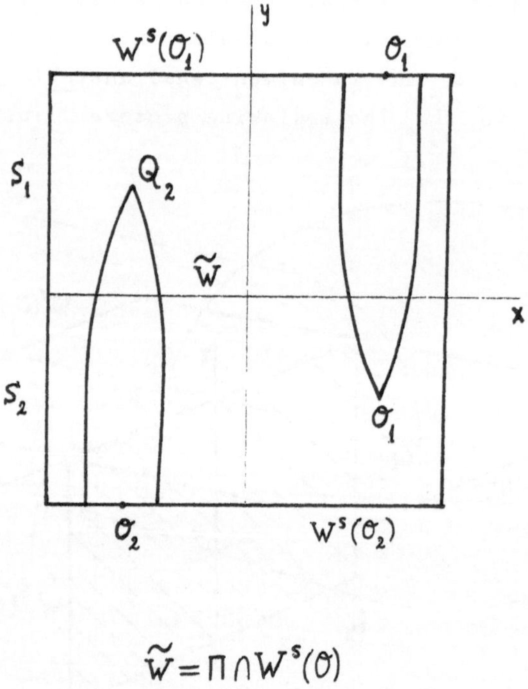

Figure 3.2.

The right hand side curvilinear triangle in the above picture is the image of the top rectangle $S_1 = \{(x,y): |x| \leq 1, 0 < y \leq 1\}$ and the left hand side triangle is the image of the bottom rectangle $S_2 = \{(x,y): |x| \leq 1, -1 \leq y < 0\}$.

Taking into account that the system (3.1) is invariant with respect to the transformation $x \to -x$, $y \to -y$, $z \to z$ we conclude that the map G can be represented by means of C^2 function $f_1(x,y)$, $g_1(x,y)$, $f_2(x,y) = -f_1(-x,-y)$, and $g_2(x,y) = -g_1(-x,-y)$ in the following way

$$G(x,y) = (f_i(x,y), g_i(x,y)) \quad \text{if} \quad (x,y) \in S_i, i = 1,2 . \qquad (3.3)$$

We assume that the functions f_1 and g_1 (and so also f_2 and g_2) can be extended continuously to $\tilde{W} = \{|x| \leq 1, \ y=0\}$ so that

$$\lim_{y \to 0} (f_i(x,y), \ g_i(x,y)) = Q_i, \ i = 1,2 . \qquad (3.4)$$

The map G has also two hyperbolic fixed points O_1 and O_2. One can proceed with the ergodic theory of the map G, as well as, of the flow F^t provided G satisfies some hyperbolicity conditions. These were not yet established rigorously. However the following conditions which according to Afraimovich, Bykov, and Shilnikov [ABS] yield the hyperbolicity of G were checked with the help of a computer by Sinai and Vul [SV]. These conditions are

$$\left\|\frac{\partial f_1}{\partial x}\right\| < 1, \ \left\|\left(\frac{\partial g_1}{\partial y}\right)^{-1}\right\| < 1,$$

$$\left\|\left(\frac{\partial g_1}{\partial y}\right)^{-1} \frac{\partial f_1}{\partial y}\right\| \cdot \left\|\frac{\partial g_1}{\partial x}\right\| < \left(1 - \left\|\frac{\partial f_1}{\partial x}\right\|\right)\left(1 - \left\|\left(\frac{\partial g_1}{\partial y}\right)^{-1}\right\|\right), \qquad (3.5)$$

$$1 - \left\|\left(\frac{\partial g_1}{\partial y}\right)^{-1}\right\| \cdot \left\|\frac{\partial f_1}{\partial x}\right\| > 2 \left(\left\|\left(\frac{\partial g_1}{\partial y}\right)^{-1}\left(\frac{\partial f_1}{\partial y}\right)\right\| \cdot \left\|\left(\frac{\partial g_1}{\partial y}\right)^{-1}\right\| \cdot \left\|\frac{\partial g_1}{\partial x}\right\|\right)^{\frac{1}{2}}$$

where $\|h(x,y)\| = \sup_{x,y \in S} |h(x,y)|$. Roughly speaking, these conditions imply that G expand distances in the vertical direction in Figure 3.1 and contracts in the horizontal direction.

The Lorenz attractor for G is $K = \overline{\bigcap_{0 \le n < \infty} G^n S}$ and the corresponding attractor for the flow F^t can be written in the form $\Lambda = \bigcup_{-\infty < t < \infty} F^t K$. The attractor K consists of smooth curves stretched along the y-axis. Bunimovich and Sinai [BS] showed that there exists a G-invariant probability measure ν on K which is absolutely continuous with respect to the Lebesgue measure generated by the length on smooth curves forming K. This property determines the measure ν uniquely and this measure possesses essentially the same properties as the Sinai-Bowen-Ruelle measure discussed in Section 2.5 (see Bunimovich and Sinai [BS] and Bunimovich [Bu]). The measure ν generates the unique F^t-invariant measure μ on Λ such that

$$\left. \frac{d\mu\left(\bigcup_{0 \le t \le s} F^t \Gamma\right)}{ds} \right|_{s=0} = \nu(\Gamma) \qquad (3.6)$$

for any Borel subset $\Gamma \subset K$ whose distance from $W^s(O_1) \cup W^s(O_2)$ is positive.

Theorem 3.1. *Suppose that the conditions* (3.5) *hold true. Then invariant measures* μ^ε *of diffusion random perturbations* X_t^ε *weakly converge to* μ *as* $\varepsilon \to 0$.

The proof of this result proceeds by the method of Chapter II modified in the same way as in Section 3.1. In place of processes X_t^ε we shall consider Markov chains $Y_n^\varepsilon = X_{nr}^\varepsilon$, $n=0,1,2,\cdots$ for some small but fixed $r > 0$ which are random perturbations of the diffeomorphism $F = F^r$. The number r is chosen so that an application of F^t with $|t| \le r$ does not destroy expanding and contracting properties of the map G in the transverse and parallel to $W^s(O)$ directions, respectively. Next, if we consider the time $n = n(\varepsilon)$ of order $(\log \varepsilon)^2$ then according to Corollary II.1.1 we may restrict our attention to a neighborhood U of the attractor Λ.

Lemma 3.1. *Let U be a sufficiently small neighborhood of the attractor Λ. There exists a constant $C > 0$ such that if $x_0, x_1, \cdots, x_n$ is a δ-pseudo-orbit of $F = F^1$ staying in U and satisfying*

$$\min_{0 \le i \le n} \operatorname{dist}(x_i, W^s(0)) > C\delta \tag{3.7}$$

then one can find a point $y \in U$ such that

$$\max_{0 \le i \le n} \operatorname{dist}(x_i, F^i y) \le Cn\delta \tag{3.8}$$

where $W^s(0)$ in (3.7) denotes a connected component of the stable manifold of 0 in U containing 0 and dist here is the Euclidean distance.

In order to obtain this kind of the shadowing one combines the arguments leading to the shadowing for the hyperbolic transformation G together with the corresponding arguments valid in a neighborhood of the hyperbolic fixed point 0. Namely, in the ρ-neighborhood of $W^s(0)$ with $\rho > 0$ small but fixed the expanding and contracting in transverse and parallel to $W^s(0)$ directions, respectively, is due to the presence of the hyperbolic fixed point 0. Thus if the orbit of the flow starts in the ρ/N-neighborhood of $W^s(0)$ with N large enough and exits from the ρ-neighborhood of $W^s(0)$ then expanding and contracting will be already accumulated enough not to be destroyed until the orbit pierces S. For orbits staying outside the ρ/N-neighborhood of $W^s(0)$ we derive expanding and contracting properties along them from the corresponding hyperbolicity properties of G which follow from (3.5). The condition (3.7) enables us to avoid difficulties connected with the discontinuity of G.

Next, employing the above arguments we derive similarly to Lemma 1.2 the absolute continuity in the unstable direction of probabilities that $Y_n^\epsilon = X_{nr}^\epsilon$ arrives to a set for $n(\epsilon) \sim (\log \epsilon)^2$ steps along paths which do not approach $W^s(0)$ closer than $\epsilon^{1-\gamma}$. Since the flow F^t stretches in the transverse to $W^s(0)$ direction then in the same way as in Lemma 1.3 we conclude that for $n \geq \log(\frac{1}{\epsilon})$ and $\gamma > 0$ small enough Y_n^ϵ may belong to the $\epsilon^{1-\gamma}$-neighborhood of $W^s(0)$ with probability not exceeding ϵ^γ. After that we complete the proof of Theorem 3.1 in the same way as the proof of Theorem 1.1. We note that the technical prerequisites for our method similar to ones collected in Proposition II.3.6 can be found in Bunimovich and Sinai [BS] or easily derived from their arguments.

Remark 3.1. One can generalize this approach in order to apply the method to situations where some kind of hyperbolicity conditions holds true only for an appropriate return map of a flow and not for the flow itself.

Bibliography

[ABS] V.S. Afraimovich, V.V. Bykov, and L.P. Shilnikov, On structurally unstable attracting limit sets of Lorenz attractor type, Trans. Moscow Math. Soc. 1983, Issue 2, 153-216.

[Ar1] D.G. Aronson, The fundamental solution of a linear parabolic equation containing a small parameter, Illinois J. Math. 3 (1959), 580-619.

[Ar 2] D.G. Aronson, Non-negative solutions of linear parabolic equations, Ann. Scuola Norm. Sup. Pisa (3) 22 (1968), 607-697.

[Bi] P.Billingsley, Convergence of Probability Measures, J. Wiley, New York, 1968.

[BK] M. Brin and Yu. Kifer, Dynamics of Markov chains and stable manifolds for random diffeomorphisms, Ergodic Th. and Dynam. Syst. 7(1987), 351-374.

[Bl] M.L. Blank, Stochastic attractors and their small perturbations, in: Mathematical Problems of Statistical Mechanics and Dynamics, ed. R.L. Dobrushin, D. Reidel, Dordrecht, 1986, 161-197.

[Bow] R. Bowen, Equilibrium States and the ergodic Theory of Anosov Diffeomorphisms, Lecture Notes in Math. 470, Springer-Verlag, Berlin, 1975.

[Boy] A. Boyarsky, Randomness implies order, J. Math. Analysis Appl. 76, (1980), 483-497.

[BR] R. Bowen and D. Ruelle, The ergodic theory of Axiom A flows, Invent. Math. 29 (1975), 181-202.

[BS] L.A. Bunimovich and Ya.G.Sinai, Stochasticity of the attractor in the Lorenz model, in: Nonlinear Waves, ed. A.V. Gaponov-Grekhov, Nauka, Moscow, 1979, 212-226 (in Russian).

[Bu] L.A. Bunimovich, Statistical properties of Lorenz attractors, in: Nonlinear Dynamics and Turbulence, ed. G.I. Barenblatt et al., Pitman, Boston, 1983, 71-92.

[CE1] P. Collet and J.-P. Eckmann, Iterated Maps of the Interval as Dynamical Systems, Birkauser, Boston, 1980.

[CE2] P. Collet and J.-P. Eckmann, Positive Lyapunov exponents and absolute continuity for maps of the interval, Ergodic Th. and Dynam. Syst. 3, (1983), 13-46.

[CFS] I.P. Cornfeld, S.V. Fomin, and Ya.G. Sinai, Ergodic Theory, Springer-Verlag, Berlin, 1982.

[Col] P. Collet, Ergodic properties of some unimodal mappings of the interval, Preprint, Inst. Mittag-Leffler, 1984.

[Con] C. Conley, Isolated Invariant Sets and the Morse Index, CBMS Regional Conference, Ser. 38, Amer. Math. Soc., Providence RI, 1978.

[Do] Y.L. Doob, Stochastic Processes, J. Wiley, New York, 1953.

[Db] N.G. DeBruijn, Asymptotic methods in analysis, Dover, New York, 1981.

[Ei] A. Eizenberg, The exit distributions for small random perturbations of dynamical systems with a repulsive type stationary point, Stochastics 12 (1984), 251-275.

[EK] A. Eizenberg and Yu. Kifer, The asymptotic behavior of the principal eigenvalue in a singular perturbation problem with invariant boundaries, Probability Theory and Related Fields, 76 (1987), 439-476.

[Fre] M.I. Freidlin, The averaging principle and theorems on large deviations, Russian Math. Surveys 33 No. 5 (1978), 117-176.

[Fri 1] A. Friedman, Parial Differential Equations of Parabolic Type, Prentice-Hall, Englewood Cliffs, New York, 1964.

[Fri 2] A. Friedman, Stochastic Differential Equations and Applications, vols. 1 and 2, Academic Press, New York, 1975.

[FW] M.I. Freidlin and A.D. Wentzell, Random Perturbations of Dynamical Systems, Springer-Verlag, New York, 1984.

[GH] J. Guckenheimer and P. Holmes, Nonlinear Oscillations, Dynamical Systems, and Bifurcations of Vector Fields, Springer-Verlag, New York, 1983.

[Gol] A.O. Golosov, Small random perturbations of dynamical systems, Trans. Moscow Math. Soc. 1984, Issue 2, 251-271.

[Gor] P. Gora, Random composing of mappings, small stochastic perturbations and attractors, Z. Wahrscheinlichkeitstheorie verw. Gebiete 69 (1985), 137-160.

[Ha] P.R. Halmos, Measure Theory, Springer-Verlag, New York, 1974.

[Has] R.Z. Hasminskii, Stochastic stability of differential equations, Sijthoff & Noordhoff, Alphen aan der Rijn, 1980.

[HPS] M.W.Hirsch, C.C. Pugh, and M. Shub, Invariant Manifolds, Lecture Notes in Math. 583, Springer-Verlag, Berlin, 1977.

[Hu] M. Hurley, Attractors: persistence, and density of their basins, Trans. Amer. Math. Soc. 269 (1982), 247-271.

[IW] N. Ikeda and S. Watanabe, Stochastic Differential Equations and Diffusion Processes, North-Holland/ Kodansha, Amsterdam, 1981.

[Ka] A. Katok, Bernoulli diffeomorphisms on surfaces, Ann. of Math. 110 (1979), 529-547.

[Ke] J.L. Kelley, General Topology, Van Nostrand, Princeton, 1957.

[Kh] R.Z. Khasminskii, The averaging principle for parabolic and elliptic differential equations and Markov processes with small diffusion, Theor. Probability Appl. 8 (1963), 1-21.

[Ki1] Yu.I. Kifer, On small random perturbations of some smooth dynamical systems, Math. USSR-Izv. 8 (1974), 1083-1107.

[Ki2] Yu.I. Kifer, On small random perturbations of diffeomorphisms, Uspehi Mat. Nauk 24 No. 4 (1974), 173-174 (in Russian).

[Ki3] Yu.I. Kifer, On the asymptotics of the transition density of process with small diffusion, Theor. Probability Appl. 21 (1976), 513-522.

[Ki4] Yu.I. Kifer, The spectrum of small random perturbations of dynamical systems, in: Multicomponent Random Systems, ed. R.L. Dobrushin and Ya. G. Sinai, Advances in Probability and Related Topics G, Marcel Dekker, New York, 1980, 423-450.

[Ki5] Yu. Kifer, On the principal eigenvalue in a singular perturbation problem with hyperbolic limit points and circles, J. Diff. Equations 37 (1980), 108-139.

[Ki6] Yu. Kifer, Stochastic stability of the topological pressure, J. D'Analyse Math. 38 (1980), 255-286.

[Ki7] Yu. Kifer, The exit problem for small random perturbations of dynamical systems with a hyperbolic fixed point, Israel J. Math. 40 (1981), 74-96.

[Ki8] Yu. Kifer, The inverse problem for small random perturbations of dynamical systems, Israel J. Math. 40 (1981), 165-174.

[Ki9] Yu. Kifer, Ergodic Theory of Random Transformations, Birkhauser, Boston, 1986.

[Ki10] Yu. Kifer, General random perturbations of hyperbolic and expanding transformations, J. D'Analyse Math. 47 (1986), 111-150.

[KK] A. Katok and Y. Kifer, Random perturbations of transformations of an interval, J. D'Analyse Math. 47 (1986), 193-237.

[Kr] M.A. Krasnoselskii, Positive Solutions of Operator Equations, Noordhoff, Groningen, 1964.

[KS] K. Krzyzewski and W.Szlenk, On invariant measures for expanding differentiable mappings, Studia Math. 33 (1969), 83-92.

[LaY] A. Lasota and J.A. Yorke, On the existence of invariant measures for piecewise monotonic transformations, Trans. Amer. Math. Soc. 186 (1973), 481-488.

[Le1] F. Ledrappier, Quelques properties des exposants characteristiques, Ecole d'Ete de Probabilitee de Saint-Flour XII - 1982, Lecture Notes in Math. 1097, Springer Verlag, 1984, 305-396.

[Le2] F. Ledrappier, Proprietes ergodique des mesures de Sinai, Publ. Math. I.H.E.S. 59 (1984), 163-188.

[LeY] F. Ledrappier and L.S. Young, The metric entropy of diffeomorphisms, Ann. of Math. 122 (1985), 509-574.

[Li] D.A. Lind, Spectral invariants in smooth ergodic theory, Lecture Notes in Phis. 38, Springer-Verlag, 1975, 296-308.

[LiY] T.Y. Li and J.A. Yorke, Ergodic transformations from an interval into itself, Trans. Amer. Math. Soc. 235 (1978), 183-192.

[LS] F. Ledrappier and J.-M. Strelcyn, A proof of the estimation from below in Pesin entropy formula, Ergod. Th. and Dynam. Syst. 2 (1982), 203-219.

[Lo] E.N. Lorenz, Deterministic non-periodic flows, J. Atmos. Sci. 20 (1963), 130-141.

[Mi] M. Misiurewicz, Absolutely continuous measures for certain maps of an interval, Publ. Math. I.H.E.S. 53 (1981), 17-51.

[Ne] J. Neveu, Mathematical foundations of the calculus of probability, Holden-Day, London, 1965.

[OW] D.S. Ornstein and B. Weiss, Statistical stability, Preprint, 1987.

[Pa] W. Parry, Entropy and generators in ergodic theory, Benjamin, New York, 1969.

[PAV] L.S. Pontryagin, A.A. Andronov, and A.A. Vitt, On statistical consideration of dynamical systems, J. Experiment. Theor. Phys. 3 (1933), No. 3, 165-180 (in Russian).

[PS] Ya. B. Pesin and Ya. G. Sinai, Gibbs measures for partially hyperbolic attractors, Ergod, Th. and Dynam. Syst. 2 (1982), 417-438.

[PW] M.H. Protter and H.F. Weinberger, Maximum principles in differential equations, Springer-Verlag, New York 1984.

[Ro] M. Rosenblatt, Markov processes, Structure and asymptotic behavior, Springer-Verlag, Berlin, 1971.

[Ru1] D. Ruelle, A measure associated with Axiom A attractors, Amer. J. Math. 98 (1976), 619-654.

[Ru2] D. Ruelle, Thermodynamic Formulism, Encyclopedia of Math. and its Appl., vol. 5, Addison-Wesley, Reading, Mass., 1978.

[Ru3] D. Ruelle, Sensitive dependence on initial condition and turbulent behavior of dynamical systems, Ann. N.Y. Acad. Sci. 316 (1978), 408-416.

[Ru4] D. Ruelle, Ergodic theory of differentiable dynamical systems, Publ. Math. 50 (1979), 275-306.

[Ru5] D. Ruelle, Small random perturbations of dynamical systems and the definition of attractors, Comm. Math. Phys. 82 (1981), 137-151.

[Ru6] D. Ruelle, Differential dynamical systems and the problem of turbulence, Bull. Amer. Math. Soc. 5, No. 1 (1981), 29-42.

[Ru7] D. Ruelle, Turbulent dynamical systems, Proc. Int. Congress of Math., Warszawa, 1983, 271-286.

[Sh] M. Shub, Global Stability of Dynamical Systems, Springer-Verlag, New York, 1987.

[Si1] Ja.G. Sinai, Gibbs measures in ergodic theory, Russian Math. Surveys 27 No.4 (1972), 21-70.

[Si2] Ja.G. Sinai, Stochasticity of dynamical systems, in: Nonlinear Waves, ed. A.V. Gaponov-Grekhov, Nauka, Moscow, 1979, 192-212 (in Russian).

[Sm] S. Smale, Differentiable dynamical systems, Bull. Amer. Math. Soc. 73 (1967), 747-817.

[Sp] C.Sparrow, The Lorenz Equations: Bifurcations, Chaos, and Strange Attractors, Springer-Verlag, New York, 1982.

[SV] Ja.G. Sinai and E.B. Vul, Hyperbolicity conditions for the Lorenz model, Physica 2D, (1980), 3-7.

[Sz] W. Szlenk, An Introduction to the Theory of Smooth Dynamical Systems, J. Wiley, Chichester and PWN, Warszawa, 1984.

[Wa] P. Walters, An Introduction to Ergodic Theory, Springer-Verlag, New York, 1982.

[WF] A.D. Wentzell and M.I. Freidlin, On small random perturbations of dynamical systems, Russian Math. Surveys 25 No. 1 (1970), 1-56.

[Yos] K. Yosida, Functional Analysis, Springer-Verlag, Berlin, 1965.

[You] L.S. Young, Stochastic stability of hyperbolic attractors, Ergod. Th. and Dynam. Syst. 6 (1986), 311-319.

Index

Anosov diffeomorphism 125

attractor 43,44

- hyperbolic - 136
- partially hyperbolic - 160
- stochastic - 164

Chapman - Kolmogorov formula 20

continuous spectrum 247

diffusion process 86,98,100,198

Doeblin's condition 27

elliptic operator 86,100

entropy 30,31

equivalence class 44,57,212

expanding transformation 123

expansive homeomorphism 31

flow 14,39,99

Frobenius-Peron operator 254

hyperbolic set 125,130

- locally maximal - 128

invariant measure

- - of a transformation 8
- - of a flow (semiflow) 14
- - of a Markov chain 8
- - of a Markov process 14

invariant set 39,40

Laplace-Beltrami operator 100,198

local product structure 128

Markov property 42

Markov chain 7

Markov process 13

partition 30

pure-point spectrum 231

pseudo-orbit 43
principal eigenvalue 200
quasiattractor 44
    ρ - 58
random perturbations 7,13
    vague - 9
    diffusion type - 86,101

Ruelle's inequality 156,159
semiflow 14
(ρ,n) - separated set 167
shadowing property 103

Sinai-Bowen-Ruelle measure 155,157
stable
    - subbundle 125
    - submanifold 127
stochastic differential equations 98,198
tight family 18
topological pressure 161,178
transition probability 7,13
uniquely ergodic 14
unstable
    - subbundle 125
    - submanifold 127
wandering point 50

Progress in Probability and Statistics

1 ÇINLAR/CHUNG/GETOOR. Seminar on Stochastic Processes, 1981
ISBN 3-7643-3072-4
2 KESTEN. Percolation Theory for Mathematicians
ISBN 3-7643-3107-0
3 ASMUSSEN/HERING. Branching Processes
ISBN 3-7643-3122-4
4 CHUNG/WILLIAMS. Introduction to Stochastic Integration
ISBN 3-7643-3117-8
5 ÇINLAR/CHUNG/GETOOR. Seminar on Stochastic Processes, 1982
ISBN 3-7643-3131-3
6 BLOOMFIELD/STEIGER. Least Absolute Deviation
ISBN 3-7643-3157-7
7 ÇINLAR/CHUNG/GETOOR. Seminar on Stochastic Processes, 1983
ISBN 3-7643-3293-X
8 BOUGEROL/LACROIX. Products of Random Matrices with Application to Schrodinger Operator
ISBN 3-7643-3324-3
9 ÇINLAR/CHUNG/GETOOR. Seminar on Stochastic Processes, 1984
ISBN 3-7643-3327-8
10 KIFER. Ergodic Theory of Random Transformations
ISBN 3-7643-3319-7
11 EBERLEIN/TAQQU. Dependence in Probability and Statistics
ISBN 3-7643-3323-5
12 ÇINLAR/CHUNG/GETOOR. Seminar on Stochastic Processes, 1985
ISBN 3-7643-3331-6
13 ÇINLAR/CHUNG/GETOOR/GLOVER. Seminar on Stochastic Processes, 1986
ISBN 3-7643-3353-7
14 DEVROYE. A Course in Density Estimation
ISBN 3-7643-3365-0
15 ÇINLAR/CHUNG/GETOOR/GLOVER. Seminar on Stochastic Processes
ISBN 3-7643-3381-2
16 KIFER. Random Perturbations of Dynamical Systems
ISBN 3-7643-3384-7